T0328160

MACHINE LEARNING AND DATA SCIENCE IN THE POWER GENERATION INDUSTRY

Dedicated to my wife Fahmida Ahmed Bangert
whose patience and understanding enabled this book to be written.

MACHINE LEARNING AND DATA SCIENCE IN THE POWER GENERATION INDUSTRY

Best Practices, Tools, and Case Studies

Edited by

PATRICK BANGERT

Artificial Intelligence, Samsung SDSA,
San Jose, CA, United States

Algorithmica Technologies GmbH,
Bad Nauheim, Germany

ELSEVIER

Elsevier
Radarweg 29, PO Box 211, 1000 AE Amsterdam, Netherlands
The Boulevard, Langford Lane, Kidlington, Oxford OX5 1GB, United Kingdom
50 Hampshire Street, 5th Floor, Cambridge, MA 02139, United States

© 2021 Elsevier Inc. All rights reserved

No part of this publication may be reproduced or transmitted in any form or by any means, electronic or
mechanical, including photocopying, recording, or any information storage and retrieval system,
withoutpermission in writing from the publisher. Details on how to seek permission, further information
about the Publisher's permissions policies and our arrangements with organizations such as the
Copyright Clearance Center and the Copyright Licensing Agency, can be found at our website:
www.elsevier.com/permissions.

This book and the individual contributions contained in it are protected under copyright by the Publisher
(other than as may be noted herein).

Notices
Knowledge and best practice in this field are constantly changing. As new research and experience broaden our
understanding, changes in research methods, professional practices, or medical treatment may become
necessary.

Practitioners and researchers must always rely on their own experience and knowledge in evaluating and using
any information, methods, compounds, or experiments described herein. In using such information or methods
they should be mindful of their own safety and the safety of others, including parties for whom they have a
professional responsibility.

To the fullest extent of the law, neither the Publisher nor the authors, contributors, or editors, assume any liability
for any injury and/or damage to persons or property as a matter of products liability, negligence or otherwise, or
from any use or operation of any methods, products, instructions, or ideas contained in the material herein.

Library of Congress Cataloging-in-Publication Data
A catalog record for this book is available from the Library of Congress

British Library Cataloguing-in-Publication Data
A catalogue record for this book is available from the British Library

ISBN: 978-0-12-819742-4

For information on all Elsevier publications
visit our website at https://www.elsevier.com/books-and-journals

Publisher: Brian Romer
Acquisitions Editor: Graham Nisbet
Editorial Project Manager: Chiara Giglio
Production Project Manager: Nirmala Arumugam
Cover Designer: Miles Hitchen

Typeset by SPi Global, India

Contents

Contributors

A. Bagnasco Department of Electrical, Electronic, Telecommunication Engineering and Naval Architecture (DITEN), University of Genova; IESolutions Soluzioni Intelligenti per l'Energia, Genova, Italy

Patrick Bangert Artificial Intelligence, Samsung SDSA, San Jose, CA, United States; Algorithmica Technologies GmbH, Bad Nauheim, Germany

Daniel Brenner Weidmüller Monitoring Systems GmbH, Dresden, Germany

Cristina Cornaro Department of Enterprise Engineering; CHOSE, University of Rome Tor Vergata, Rome, Italy

Jim Crompton Colorado School of Mines, Golden, CO, United States

Peter Dabrowski Digitalization at Wintershall Dea, Hamburg, Germany

F. Fresi Gruppo Humanitas, Clinica Cellini, Torino, Italy

Robert Maglalang Value Chain Optimization at Phillips 66, Houston, TX, United States

David Moser EURAC Research, Bolzano, Italy

Stewart Nicholson Primex Process Specialists, Warrington, PA, United States

Marco Pierro Department of Enterprise Engineering, University of Rome Tor Vergata, Rome; EURAC Research, Bolzano, Italy

M. Saviozzi Department of Electrical, Electronic, Telecommunication Engineering and Naval Architecture (DITEN), University of Genova, Genova, Italy

F. Silvestro Department of Electrical, Electronic, Telecommunication Engineering and Naval Architecture (DITEN), University of Genova, Genova, Italy

Dietmar Tilch ZF Friedrichshafen AG, Lohr am Main, Germany

A. Vinci IESolutions Soluzioni Intelligenti per l'Energia, Genova, Italy

Foreword

The power generation industry is a complex network of power plants running on various fuels, grids distributing the electricity, and a host of consumers. Many challenges exist from predicting physical events with equipment or the availability of fuels such as solar radiation or wind. They also range over issues of balancing the smart grid and negotiating electricity prices.

Many of these challenges can be solved if the underlying mechanism is exposed in numerical data and encapsulated in the language of mathematical formulas. It is the purpose of data science to obtain, clean, and curate the data necessary and the purpose of machine learning to produce the formulas. Once these exist and are verified to be correct, they are often themselves the solution or can be readily converted into an answer to the challenges.

This book will present an overview of data science and machine learning and point toward the many use cases in the power industry where these have already solved problems. The hype surrounding machine learning will hopefully be cleared up as this book focusses on what can realistically be done with existing tools.

The book is intended for any person working in the power industry who wants to understand what machine learning is and how it applies to the industry. It is also intended for machine learners to find out about the power industry and where its needs are. Finally, it is addressed to students of either subject or the general public to demonstrate the challenges faced by an industry that most are familiar with only through the power outlet on the wall, and the solutions of these challenges by methods that most are familiar with through online channels.

In combining machine learning solutions with power industry challenges, we find that it is often not technology that is the obstacle but rather managerial tasks. Project management and change management are critical elements in the workflow that enable machine learning methods to practically solve the problem. As a machine learner, it is important to pay attention to these aspects as they will decide over the success and failure of a project or tool.

This book will have served its purpose if a power company uses the advice given here in facilitating a machine learning project or toolset to drive value. It is meant partially as an instruction manual and partially as an inspiration for power company managers who want to use machine learning and artificial intelligence to improve the industry and its efficiency.

My heartfelt gratitude goes out to the coauthors of this book for being part of the journey and communicating their successes and lessons learned. Thank you to all those who have educated me over the years in the fields of power, machine learning, and management. Thank you to my wife and family for sparing me on many an evening and weekend while writing this text. Finally, thank you to you, the reader, for picking up this book and reading it! Feedback is welcome and please feel free to reach out.

Introduction

Patrick Bangert[a,b]

[a]Artificial Intelligence, Samsung SDSA, San Jose, CA, United States
[b]Algorithmica Technologies GmbH, Bad Nauheim, Germany

1.1   Who this book is for

This book will provide an overview of the field of machine learning as applied to industrial datasets in the power generation industry. It will provide enough scientific knowledge for a manager of a related project to understand what to look for and how to interpret the results. While this book will not make you into a machine learner, it will provide everything needed to talk successfully with machine learners. It will also provide many useful lessons learned in the management of such projects. As we will learn, over 90% of the total effort put into these projects is not mathematical in nature and all these aspects will be covered.

A machine learning project consists of four major elements:

1. *Management*: Defining the task, gathering the team, obtaining the budget, assessing the business value, and coordinating the other steps in the procedure.
2. *Modeling*: Collecting data, describing the problem, doing the scientific training of a model, and assessing that the model is accurate and precise.
3. *Deployment*: Integrating the model with the other infrastructure so that it can be run continuously in real time.

Machine Learning and Data Science in the Power Generation Industry. https://doi.org/10.1016/B978-0-12-819742-4.00001-9
© 2021 Elsevier Inc. All rights reserved.

4. **_Change management_**: Persuading the end-users to take heed of the new system and change their behavior accordingly.

Most books on industrial data science discuss mostly the first item. Many books on machine learning deal only with the second item. It is however the whole process that is required to create a success story. Indeed, the fourth step of change management is frequently the critical element. This book aims to discuss all four parts.

The book addresses three main groups of readers: power professionals, machine learners and data scientists, and the general public.

Power professionals such as C-level directors, plant managers, and process engineers will learn what machine learning is capable of and what benefits may be expected. You will learn what is needed to reap the rewards. This book will prepare you for a discussion with data scientists so that you know what to look for and how to judge the results.

Machine learners and **_data scientists_** will learn about the power industry and its complexities as well as the use cases that their methods can be put to in this industry. You will learn what a power professional expects to see from the technology and the final outcome. The book will put into perspective some of the issues that take center stage for data scientists, such as training time and model accuracy, and relativize these to the needs of the end-user.

For the **_general public_**, this book presents an overview of the state of the art in applying a hyped field like machine learning to a basic necessity like electricity. You will learn how both fields work and how they can work together to supply electricity reliably, safely, and with less harm to the environment.

One of the most fundamental points, to which we shall return often, is that a practical machine learning project requires far more than just machine learning. It starts with a good quality dataset and some domain knowledge, and proceeds to sufficient funding, support, and most critically change management. All these aspects will be treated so that you obtain a holistic 360-degree view of what a real industrial machine learning project looks like.

The book can be divided into two parts. The first chapters discuss general issues of machine learning and relevant management challenges. The second half focuses on practical case studies that have been carried out in real industrial plants and reports on what has been done already as well as what the field is capable of. In this context, the reader will be able to judge how much of the marketing surrounding machine learning is hype and how much is reality.

1.2 Preview of the content

The book begins in Chapter 2 with a presentation of data science that focusses on analyzing, cleaning, and preparing a dataset for machine learning. Practically speaking, this represents about 80% of the effort in any machine learning project if we do not count the change management in deploying a finished model.

We then proceed to an overview of the field of machine learning in Chapter 3. The focus will be on the central ideas of what a model is, how to make one, and how to judge if it is any good. Several types of model will be presented briefly so that one may understand some of the options and the potential uses of these models.

A review of the status of machine learning in power generation follows in Chapter 4. While we make no attempt at being complete, the chapter will cover a large array of use cases that have been investigated and provides some references for further reading. The reader will get a good idea of what is possible and what is hype.

In Chapter 5, Jim Crompton addresses how the data is obtained, transmitted, stored, and made available for analysis. These systems are complex and diverse and form the backbone of any analysis. Without proper data collection, machine learning is impossible, and this chapter discusses the status in the industry of how data is obtained and what data may be expected.

Management is concerned with the business case that Robert Maglalang analyzes in Chapter 6. Before doing a project, it is necessary to defend its cost and expected benefit. After a project, its benefit must be measured and monitored. Machine learning can deliver significant benefits if done correctly and this chapter analyzes how one might do that.

Machine learning projects must be managed by considering various factors such as domain expertise and user expectations. In a new field like machine learning, this often leads to shifting expectations during the project. In Chapter 7, Peter Dabrowski introduces the agile way of managing such projects that has had tremendous successes in delivering projects on time, in budget, and to specifications.

The next several chapters discuss concrete use cases where machine learning has made an impact in power generation.

In Chapter 8, Cristina Cornaro presents work for the electricity grid in Italy considering its photovoltaic capacity. Issues like grid stability and economic factors of pricing are analyzed to illustrate the complexity of running a grid. Forecasting weather conditions that influence the capacity of solar power is a crucial element and so we see a variety of machine learning models working together at various levels in the system to make sure that whole system works well.

The power demand of a complex building such as a hospital is presented by Andrea Bagnasco in Chapter 9. Forecasting, planning, analysis, and proper communication can lead to a reduction in power usage without sacrificing any utility in the building.

Environmental pollution such as the release of NOx or SOx gasses into the atmosphere while operating a gas turbine is harmful. With machine learning, physical pollution sensors can be substituted by models. These are not only more reliable, but they allow model predictive control and thus are able to lower pollution. Shahid Hafeez presents this in Chapter 10.

Many factors influence the efficiency of a power plant. Stewart Nicholson illustrates this in Chapter 11 and presents a way to objectify the selection of factors, their ranking, and analyzing their sensitivity. In turn, these lead to direct ways to increase the efficiency and thus lower waste.

Daniel Brenner and Dietmar Tilch talk about wind power in Chapter 12 by classifying and forecasting damage mechanisms in the most famous of all industrial machine learning use cases: predictive maintenance. They find that it is possible to accurately forecast and identify a problem several days before it takes place.

At this point, the book welcomes you to take the lessons learned and the intellectual tools acquired and put them into practice in a real machine learning project.

1.3 Power generation industry overview

Electricity is one of the most central elements of our modern civilization. It literally shines light into darkness and therefore enables education. It runs household and handheld appliances and therefore liberates much repetitive manual work in the home and in factories, which revolutionized many social norms. It forms the basis of the internet and the entertainment industry that democratize knowledge and leisure. Many more consequences to the generation and distribution of electricity could be mentioned. It is hard to imagine our world without it. As compared to several other basic utility industries, power generation is surprisingly complex.

Most methods that produce electricity do so by causing a generator to turn. The only notable exception to this is photovoltaic power, which is based on the photoelectric effect. The generator is usually turned by a turbine that converts linear motion into a turning motion. This linear motion is supplied by some fluid that we cause to move through the turbine blades. This could be *wind*, *water*, or *steam*.

Most electricity worldwide is generated by steam moving through the turbine. Steam is easily generated by heating water. The source of the heat is a fire fueled by *gas*, *coal*, *oil*, or a *nuclear fission* reaction. Oversimplifying the situation somewhat, electricity is therefore easily generated by lighting a fire underneath a vat of water and catching the steam with a turbine that turns a generator. In practice, this is complex because we need to do this safely and economically at scale. Electricity generation by gas turbine is similar but does not involve heating water as the combusted gas directly turns the turbine.

As most power is generated by first producing heat, this heat can be repurposed for secondary uses, namely district heating to heat homes or heat in the form of steam for industrial production. Many power generation facilities are therefore combined heat and power (CHP) plants that supply both electricity and heat to a neighboring industrial facility or city.

Electricity is difficult to store over time. While battery technologies exist, these usually work only for small amounts of electricity that might power a mobile phone for a day or drive a car for a few hundred miles in the best case. Research and development is ongoing to provide a battery capable of storing the output of an industrial-scale power plant but this is not economically practical at scale at this time. For the most part, the electricity that we use right now must be generated right now as well. To facilitate the distribution of electricity from the plant to the individual user without disruptions, most places have built a *power grid* that connects many producing plants and a vast number of users into a single system. For example, the European power grid connects 400 million customers in 24 countries and over 4000 generation plants, not counting the large number of households, that provide power via solar cells on their roofs, or individual wind turbines.

Unifying many kinds of power generators and a huge number of users into a single real-time cross-border system gives rise to economic problems. The demand of users for both electricity and heat fluctuates significantly over the course of a day and over the course of year due to the weather, weekends, holidays, special events, and so on. Forecasting demand is a major challenge. Prices and production volumes must be agreed upon very quickly and at frequent intervals, e.g., every 15 min.

Some renewable energy sources, first and foremost solar and wind power, depend on resources that quickly change in an uncontrollable manner. Forecasting the expected power output of such generators is a significant problem not only for their owners but for the grid's stability, as input and output at any one time must be (roughly) the same. Some nations have enacted laws that give priority to these power sources meaning that other power generators must either fill in the gaps or cycle down in order not to overload the grid. This puts significant pressure on conventional energy sources to be flexible with assets that were not designed for it.

The power generation industry is a complex industry that has diverse data-driven challenges due to the decentralized nature of providing electricity to every office and home. It starts with providing fuel to plants and running large capital equipment. That power needs to be distributed through the grid and transformed to the right voltage all the way to the end-user. The by-product heat needs to be distributed wherever this makes sense. The undesirable by-products such as CO_2 and other pollutants must be detected and removed in suitable ways. Many providers and users must be unified in an economic system that is sustainable at each moment requiring reliable forecasts. This book aims to study this complex system by using machine learning and illustrating several lighthouse applications.

1.4 Fuels as limited resources

The three *fossil fuels* of gas, coal, and oil make up 64% of worldwide electricity generation with *nuclear* providing another 10%. The *renewable sources* of hydropower (16%), wind (4%), solar (2%), and biofuels (2%) make up 24%; the remaining 2% are diverse special sources.[a]

That means that 74% of the electricity generated today is generated from fuels that are finite resources that eventually will be used up. Of course, these resources are being replenished even now but the making of gas, coal, and oil requires millions of years while the making of uranium (the fuel for nuclear fission) cannot be made on Earth at all as it requires a supernova.

The Club of Rome published a famous study in 1972 entitled "The Limits to Growth" that analyzed the dependency of the world upon finite resources. While the chronological predictions made in this report have turned out to be overly pessimistic, their fundamental conclusions remain valid, albeit at a future time

[a] According to the International Energy Agency in 2017.

(Meadows et al., 1972). A much-neglected condition was made clear in the report: The predictions made were made considering the technology available at that time, i.e. the prediction could be extended by technological innovation. This is, of course, exactly what happened as further analyzed in the update of the report (Meadows et al., 2004). Humanity innovated itself a postponement of the deadline. The problem remains, however. Growth cannot occur indefinitely and especially not based on resources that are finite. The ultimate finite resource is land, but this is quickly followed by gas, coal, oil, and uranium, considering their places in our world.

The global climate is changing at a fast rate. This fact is largely related to the burning of fossil fuels by humanity since the start of the industrial revolution. Considering the present-day problem of climate change, the finite nature of these fuels may be a theoretical problem—it is likely that the consequences of climate change will become significant for the ordinary individual far earlier than the consequences of reduced fuel availability.

Both problems, however, have the same solution: We must find a way to live without consuming as many resources. We must do this in part because they are finite and in part because their use is harmful. This new way is likely to involve two major elements: technological innovation and lifestyle change.

Technological innovations will be necessary to overcome some fundamental challenges such as how wasteful the process is from the raw fuel to the electricity in the home. As this is a scientific challenge, we must figure out how complex processes work and how we can influence them. This is the heart of ***data science*** with ***machine learning*** as its primary toolbox.

Renewable energy sources (such as wind, solar, hydro, biomass, and others) are examples of technological innovations that already exist. Many predictions are made about their future and we cannot be certain about any of them. There is significant uncertainty as to how much electricity can be generated this way once these technologies are deployed at scale. The primary factors are the amount of land required, which competes with agriculture and living spaces, as well as the amount of materials needed to build and maintain the devices, e.g., a wind turbine or a solar panel. The largest uncertainty for the system is the intermittent availability that must be buffered with electricity storage methods that do not exist yet. If you are looking to make your mark in power generation, this would be it: Design a battery that can store large amounts of electricity and can be manufactured at scale without running into other finite resource problems, e.g. the availability of lithium.

Lifestyle change will be needed from every individual on the planet in that we must use our resources with more care—essentially, we must use far fewer resources. This will involve effort on an individual level, but it will also involve a transformation of societies. Public transportation must supplant individual transportation. Consumerism must be overcome, and products must be of better quality to last longer. Society must rely more on society, rather than material objects. To transform society in a manner that most people are happy with the transition requires careful analysis. In large numbers, humans are predictable and amenable to numeric description and study. It is again the field of data science and machine learning that is pivotal in enabling the transformation through insight and design of the best systems for the future.

1.5 Challenges of power generation

These issues provide a range of challenges to the power generation industry worldwide. We must continue to find and produce gas, coal, oil, and uranium during the transition to a new social norm. The process must become more efficient and less wasteful. There are many dangers along the way that must be mitigated. This starts from injury to individual workers and goes to global threats as illustrated by the Deepwater Horizon disaster in 2010. The *digital transformation* is a journey poised to solve these challenges. It promises to do this by increasing our understanding and control of the process. At the beginning are sensors and communication software. At the end are physical technologies like drones, 3D printing, autonomous vehicles, or augmented reality.

At one end of the spectrum are all the technologies needed to find and produce the fuel sources, e.g. mining. These fuels are transported to power generation plants where they are turned into electricity. These plants operate many large machines such as pumps, compressors, turbines, generators, transformers, and more. The largest cost in operating such a plant is maintenance over a lifespan of several decades. The plants are connected to a grid that is itself an asset-intensive and extremely distributed network of devices. The entire system must be regulated by balancing supply and demand. This requires a complex system of negotiation including many separate actors.

At an abstract level, most of the issues boil down to forecasting something, telling if something is as it should be or not, determining how much of something has been made, and deciding what to

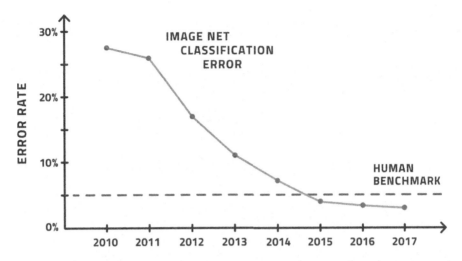

Fig. 1.1 The evolution of machine learning as compared to human performance relative to a common standard dataset.

do in response to an outside stimulus. This sounds very much like what our brain does for our body: Is it a tiger? Will it hurt? Should I run away?. Elevating the reptilian brain of the current system of power generation to an intelligent brain is the job of machine learning (ML).

ML is the piece of the puzzle that converts a dataset into a formula, also called a model. Once we have a model and we are confident that the model is right, we can do many things with it. A popular challenge these days is classifying images into various categories, see Fig. 1.1. Any individual human being correctly classifies about 95% of images and makes 5% errors. Until 2014, computers performed worse. Starting in 2015 however, computer models based on machine learning started to outperform humans for the first time. This is an example that is consistently repeated time and again with many different datasets and tasks. Machine learning models are now more accurate than humans, calculate their output in a fraction of a second, and can keep doing it with the same accuracy without limit.

As they require little human effort to make and maintain as well as being fast to compute, machine learning models are much cheaper than any alternative. These two features also enable many novel use cases and business models that simply could not be implemented until now. In relation to the power generation industry, there are six main areas that machine learning can help with, see Fig. 1.2:

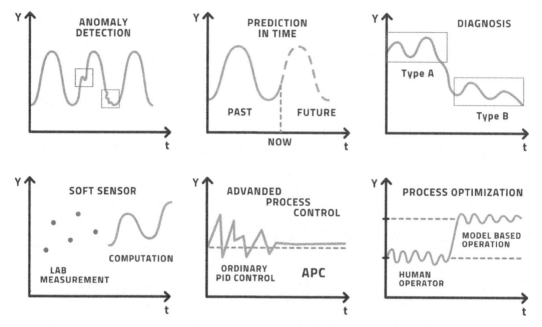

Fig. 1.2 The six main application areas of machine learning in power generation.

1. Is this operation normal or not?
2. What will the supply, demand, or operational condition be at a future time?
3. Which category does this pattern belong to?
4. Can this complex, expensive, fragile, or laboratory measurement be substituted by a calculation?
5. How shall we adjust set points in real time to keep the process stable?
6. How and when shall we change set points to improve the process according to some measure of success?

One of the biggest topics in industrial machine learning is **pre-dictive maintenance**, which is the combination of the first three points in this list. Is my equipment performing well right now, how long will that remain and when it fails, what is the damage type? Once we know a damage type and a failure time in the future, we can procure spare parts in advance and schedule a maintenance measure to take place preemptively. This will prevent the actual failure and thus prevent collateral damage, which is typically 90% of the total financial cost of a failure.

References

Meadows, D.H., Meadows, D.L., Randers, J., Behrens III, W.W., 1972. The Limits to Growth: A Report for the Club of Rome's Project on the Predicament of Mankind. Universe Books, White River Junction.

Meadows, D.H., Randers, J., Meadows, D.L., 2004. The Limits to Growth: The 30-Year Update. Chelsea Green Publishing, White River Junction.

2

Data science, statistics, and time series

Patrick Bangert[a,b]

[a]Artificial Intelligence, Samsung SDSA, San Jose, CA, United States
[b]Algorithmica Technologies GmbH, Bad Nauheim, Germany

The data that we are likely to encounter in the power generation industry is usually numerical data ordered in time. For example, a temperature that is measured every 10 min for years. The source of this data is a sensor deployed in the field. This sensor may need to be recalibrated, or repaired, or exchanged every so often but the chain of measurements continues. The collection of these measurements is called a **time series**. The label in the database that identifies this time series versus others is usually called a **tag**. For a presentation of more traditional time series analysis that does not use machine learning, see Hamilton (1994).

In any one plant, or field, we are likely to encounter tens of thousands of tags. Each of these may have a history of years. The

Machine Learning and Data Science in the Power Generation Industry. https://doi.org/10.1016/B978-0-12-819742-4.00002-0
© 2021 Elsevier Inc. All rights reserved.

frequency of records may range from one measurement a day to several measurements per second, depending on the application.

Much of the data science and machine learning in the popular literature concerns itself with data in different forms. Images, series of words, audio recordings, videos, and the like are not frequently encountered in the industry. These have become popular through a variety of consumer applications of data science and we will not discuss them here.

Apart from the time series, we encounter another type of data in the industry: the spectrum. A vibration is typically measured as the amount of movement for each of many different frequencies. The collection of these measurements is called a **vibration spectrum**. These may then be recorded at regular intervals over time, forming a time series. Normal time series are numbers, but these time series are vectors as each observation is a string of numbers. There is significant scientific analysis that deals with spectra and this can be brought to bear on these spectra as well, especially the Fourier transform (Press et al., 2007).

As we look at our data, we gradually develop more high-level understanding of what this data tells us. Fig. 2.1 illustrates the five major stages in this journey. First, all we have is a large collection of numbers, the data itself. Second, the data becomes information once all the duplicates, outliers, useless measurements, and similar are removed and we are left with relevant and significant data. Third, knowledge is gained once it is clear how the data is connected, which tags are related to other tags, how causes relate to effects, and so on. Fourth, insight is generated once we know where we are and where we need to be. Fifth, wisdom is when we know how to get to where we need to be. All that is left at this point is to execute the plan.

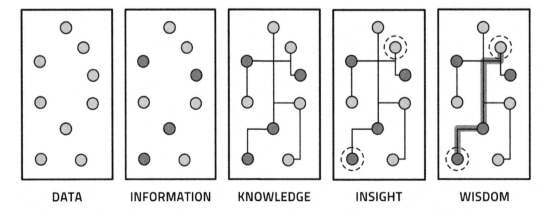

Fig. 2.1 The five qualitatively different stages of data science.

This chapter will introduce some common techniques of dealing with numerical data in preparation for machine learning (Pyle, 1999).

2.1 Measurement, uncertainty, and record keeping

Sensors measure quantities based on some sort of physical response and usually convert this into an electrical signal. This is then converted into a number using an analog-to-digital converter and sent to the **process control system**. This cycle may happen many times per second. The process control system typically stores this data only for a short while. It is the job of the **historian** to save the data for the long term and make it available for analysis, typically in diagrammatic form.

Some quantities are easy to measure, like temperature (Fig. 2.2). Some are difficult to measure, like heat lost to the environment. Finally, some quantities are impossible to measure directly and must be computed based on other measurements, like enthalpy. When measuring something, we must choose the right sensor equipment, and this must be placed correctly in order to give sensible outputs.

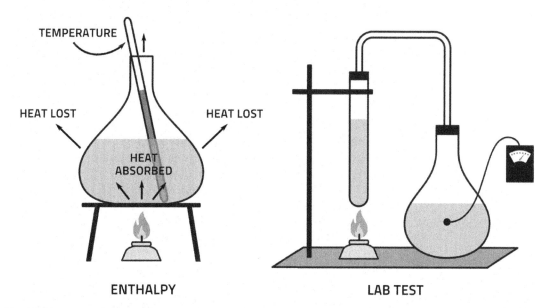

Fig. 2.2 Measurements both easy and hard.

Suppose that we have measured some temperature to be 1.5°C at time $t = 0$, 3°C at time $t = 1$, and 2.5°C at time $t = 2$. With this data at hand, we ask two crucial questions: (1) How accurate is the measurement, and (2) how shall we record the measurements in the historian?

2.1.1 Uncertainty

Every sensor has a tolerance and so does the analog-to-digital conversion. Some electrical signal loss and distortion is introduced along the way from the sensor to the converter. Sensors drift over time and must be recalibrated at regular intervals. Sensors get damaged and must be repaired. Accumulations of material may cover a sensor and distort its measurement. Numbers are stored in IT systems, such as the control systems, with finite precision and there is loss of such precision with every conversion or calculation. All these factors lead to a de facto uncertainty in any measurement. The total uncertainty in any one measurement may be difficult to assess. In the temperature example, we may decide that the uncertainty is ±1°C. This would make the last two observations—of 3 and 2.5—numerically distinct but actually equal.

Any computation made based on an uncertain input has an uncertainty of its own as a consequence. Computing this inherited uncertainty may be tricky—depending on the computation itself—and may lead to a larger uncertainty than the human designer anticipated. In general, the uncertainty in a quantity y that has been computed using several uncertain inputs x_i is the total differential,

$$(\Delta y)^2 = \sum_{i=1}^{N} \left(\frac{\partial y}{\partial x_i} \right)^2 (\Delta x_i)^2$$

In practice, this might be difficult to use and so we find that it is estimated in many cases, as illustrated in Fig. 2.3.

To anyone who has ever spent time in a laboratory, it is a painful memory that the treatment of measurement errors consumes many hours of work and is competitive in effort to the entire rest of the analysis. When engaged in data science, we must not forget that these data are not precise and therefore any conclusion based on them is not precise either. Some effects that are visible in the data quickly disappear once the uncertainty of the data is considered!

In general, it is advisable never to believe any data science conclusion in the absence of a thorough analysis of uncertainties.

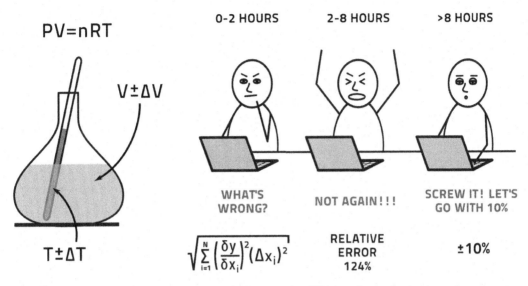

Fig. 2.3 Determining the uncertainty of a computed result can be difficult and may need to be estimated.

2.1.2   Record keeping

The historian will save the measurements to its database. The IT systems involved were often designed and deployed during a time when hard disk space was expensive and limited. There was a desire to minimize the amount of space that data takes up. An obvious idea is to only record a value if it is different from the last recorded value by more than a certain amount. This amount is known as the **compression factor** and is an essential property of the tag. Someone, usually at the time of commissioning the historian, decided on the compression factor for each tag. Frequently the default value was left in place or some relatively large value was chosen in an effort to save space. Fig. 2.4 illustrates the situation with the temperature measurements alluded to earlier and a compression factor equal to 1.

The information that is contained in the data points not recorded is lost forever. If the compression factor is lower than the uncertainty, then there is no useful information in the lost data as all we would record are random fluctuations. However, if the compression factor is larger than the uncertainty, we lose real information about the system.

It is an easy fix for any company operating a data historian to review the compression factors. In many cases, one will find that they are too large and must be lowered. This change alone will provide the operator with significantly more valuable data.

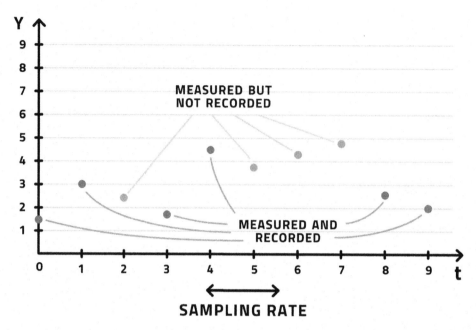

Fig. 2.4 A sensor outputs its value at every time step but not every value is recorded if the compression factor is 1.

As the compression factor will prevent the recording of some measurements, what shall we do if the historian is supposed to output the value for a time for which it has no record? See Fig. 2.5 for an illustration. The data for $t = 3$ is missing. There are several different ways to answer a query for this value:

A. Staircase: The value remains constant at the last recording until a new measurement is recorded.

B. Interpolation: Some nonlinear function is used to interpolate the data. Usually this is a spline curve.

C. Linear: A straight line is drawn in between each successive observation to interpolate the data.

While there are arguments both for and against any one of these, it is the general industry standard and agreement to use the first method—the staircase. A tag is equal to the last validly recorded value. This has the distinct advantage of holding in a real-time context, i.e., in a context where we need to infer the value at the present moment without knowing the value of the next measurement in the time series.

The correct configuration of the data historian forms the basis for the available dataset for any analysis. If we are presented with a ready-made dataset, the uncertainty of each tag is typically dominated completely by the compression factor. In practice

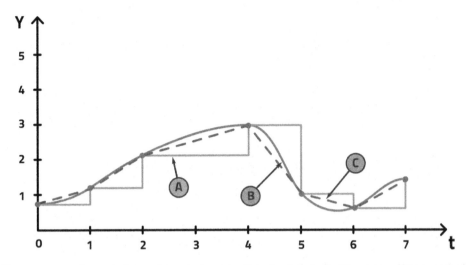

Fig. 2.5 There are several ways to interpolate missing data. Here the data for $t=3$ is missing. Which method shall we choose?

therefore, we data scientists may say that the uncertainty *is* the compression factor. Nevertheless, the compression factor should be examined by the process engineers.

2.2 Correlation and timescales

As one tag changes its value, we may be able to say something about the change in another tag's value. For example, as temperature increases, we would expect the pressure to increase also. These two tags are called **correlated**, see Fig. 2.6. The change in the first tag may provide *some* information about the change in the second tag, but not all. A measure that indicates how much information is provided is the **correlation coefficient**. This is a number between -1 and 1. Positive numbers indicate that as the first tag increases, so does the second tag. Negative numbers indicate that as the first tag increases, the second decreases. An absolute value of 1 indicates that the change happens in exact lockstep. As far as change is concerned, these two tags are identical. A value of 0 indicates that the two tags are totally unrelated. Change in the first tag provides no information about the change in the second whatsoever.

Datasets from power generation facilities concern many interconnected parts. Many tags will be correlated with each other, such as temperature and pressure at one location. Analyzing

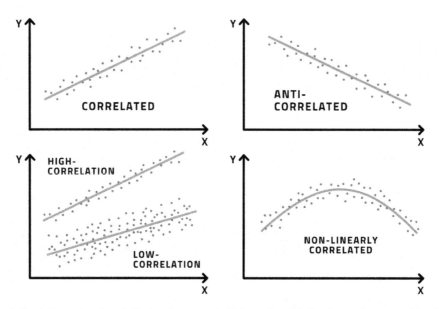

Fig. 2.6 Correlation between two tags indicates the amount of information that the change in one provides for the other.

which values are more and less correlated naturally provides an interesting clustering on the tags.

Generally, it is understood that correlation is measured in the context of linear dependency as illustrated in Fig. 2.6. The correlation coefficient due to Pearson, which is the measure usually used, indicates the strength of a linear relationship. If there is a nonlinear relationship, then this may not be picked up by the correlation coefficient in any meaningful way. This is important to take into account as many physical processes are nonlinear. Nonlinear correlation analysis presupposes that we know the kind of nonlinearity that is present, and this is another challenge. In practice, we find the linear correlation coefficient used exclusively. This is fine if it is understood properly.

As we are dealing with time series, it is important to say that correlation is typically calculated for values recorded at the same time. If we shift one of the two time series by a few time steps, then we get the **correlation function**, i.e., the correlation coefficient as a function of the **time lag**. In the special case that we compute the correlation of a tag relative to itself for different time lags, then this is called the **autocorrelation function**. These functions may look like Fig. 2.7.

We see that the correlation is high for a time lag of 0 and then decreases until it reaches a minimum at a time lag of 10 only to rise to a secondary maximum at a lag of 20, and so on. These

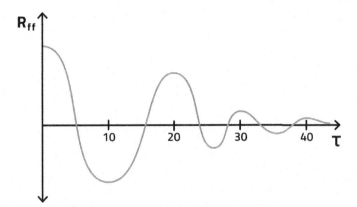

Fig. 2.7 The correlation function between two tags as a function of the time lag.

two time series are clearly related by some mechanism that takes about 10 time steps to work. If there is an amount of time that is inherent to the dynamics of the system, we call it the **timescale** of the system. It is very important to know the timescales of the various dynamics in the system under study. Some dynamics may go from cause to effect in seconds while others may take days.

It is important to know the timescales because they determine the frequency at which we must observe the system. For example, in Fig. 2.7 the timescale is clearly 10. If we observe the system once every 20 time steps, then we lose important information about the dynamics of the system. Any modeling attempt on this dataset is doomed to fail.

A famous result in the field of signal processing is known as the **Nyquist-Shannon sampling theorem** and it says, roughly speaking: If you want to be able to reconstruct the dynamics of a system based on discrete observations, the frequency of observations must be at least two observations for the shortest timescale present in the system. This is assuming that you know what the shortest timescale is and that the data is more or less perfect. In industrial reality, this is not true. Therefore in practical machine learning, there is a rule of thumb that we should try for 10 observations relative to the timescale of the system. In the example of Fig. 2.7, where we measured the system every 1 time step, the observation frequency was therefore chosen well.

The other reason for choosing to sample more frequently than strictly necessary is an effect known as **aliasing** sometimes seen in digital photography. This is illustrated in Fig. 2.8. Based on multiple observations, we desire to reconstruct the dynamics and there are multiple candidates if the sampling frequency is tuned perfectly to the internal timescale. The danger is that the wrong one is chosen, resulting in poor performance. If we measure more frequently, this problem is overcome.

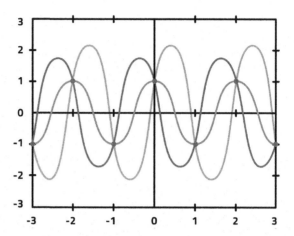

Fig. 2.8 The aliasing problem in that the observations allow multiple valid reconstructions.

In Fig. 2.8, the dots are the observations and the lines are various possible reconstructions of the system. All reconstructions fit the data perfectly but are quite different from each other. The data is therefore not sufficient to characterize the system properly and we must make an arbitrary choice.

The practical recommendation to a data scientist is to obtain domain knowledge on the system in question in order to determine the best sampling frequency for the task. Sampling the data less frequently than indicated earlier results in loss of information. Sampling it more frequently will only cost the analysis in time and effort. On balance, we should err on the side of measuring somewhat more frequently than deemed necessary. When working with transient systems, the right frequency might change depending on the present regime. In those cases, we will want to choose the fastest frequency from the entire system.

2.3 The idea of a model

Physical phenomena are described by physics using the language of mathematics. The expression of this insight is an equation that allows us to predict the behavior of physical systems and thus establish control over them. Take, as an example, the ideal gas law. It states that an ideal gas of n moles that is contained in a volume V, under pressure P, and at temperature T obeys the relationship

$$PV = nRT$$

where R is a constant. This relationship is simple to understand and allows us to exert control over the gas by changing one of the three variables in order to change the others. For example, if the volume remains the same and we increase the temperature, the pressure will rise.

This equation is an example of a **model**. A model is an equation that allows us to compute a quantity of interest from other quantities that we can obtain in some way, usually by measurement. The model allows us to calculate this quantity instead of measuring it in the real world. On the one hand, a model saves us the effort and time of measurement and verification. On the other hand, it lets us know in advance what will happen and therefore choose our actions according to what we want to happen. Finally, a model like this also provides understanding.

This model was discovered in 1834 by Émile Clapeyron based on the previous work of several other scientists who had gone through a time-consuming process of performing many physical experiments and carefully measuring all the related quantities. Detailed plots were created by hand and studied until the ideas emerged one by one of the linear relationships between the three state variables and the amount of gas present. Finally, all these partial results were combined into the result—the ideal gas law.

The model was produced by two major factors: patient collection of data and human learning.

Let us say, for the time being, that all we had was a large table of numbers with one column each for volume, pressure, temperature, and amount of gas. We suspect that there is some relationship between these four concepts, but we do not know what it is. What we want is a model that can calculate the pressure from the other three quantities.

Today, what we would do is submit this table of numbers to a computer program that uses techniques called **machine learning**. This computer program would make a model and present some results to us that compare the pressure column from the table and the pressure as obtained from the proposed model. In each experimental case, i.e., for each row in the data table, we have these two pressures: the measured and the calculated pressure. If the model is good, these two are always very close to each other. In addition, we usually want the model to be relatively simple in order to overcome some potential problems that we will go into later.

If we are satisfied with this analysis, we may conclude that the model represents the situation sufficiently well to be useful. We have a model that looks like

$$P = f(n, V, T)$$

The function $f(\cdots)$ is not usually something that makes sense to write down on paper, to look at with our eyes, and to try to understand. In most cases, the function $f(\cdots)$ is some large matrix of numbers that is unreadable. We must sacrifice understanding for this model. However, we get a model without having to do the learning ourselves in our head. The more complex and larger the data, the more return we get for our payment in understanding.

Despite being unintelligible—pardon the pun—the model is useful in that it is computable. We can now calculate the pressure from the other quantities. Having performed the analysis of the differences between computed and measured pressures over many known pressures in diverse situations, we are confident that the model produces the correct numerical value and that is all we can expect.

In the modern day, our datasets tend to be so large, the number of variables so many, and the situations so complex, that a manually made model based on human insight is often not feasible, economical, or even possible. In many situations, we already possess the basic understanding of the situation, e.g., if this increases, then that increases also, but we need a formula to tell us quantitatively how large the changes are exactly.

This insight is comparable to a long-standing debate in physics: Is it the job of physics to explain why something is happening or simply to correctly predict what will happen? Where you stand on this idea is the watershed for machine learning as opposed to physical modeling. If you are prepared to sacrifice understanding to some degree, you can gain tremendous performance at the same time as a huge reduction in cost.

2.4 First principles models

Many models, especially those for physical phenomena, can be derived from the laws of physics. Ultimately, this means that we have a set of coupled partial differential equations that probably combine a host of characteristics of the physical system that we must know and provide to the model. We might have to know what material every component is made from, how large every piece is, and how it was put together. Constructing such a model can be a daunting task requiring multiple experts to spend months of their time.

Solving this set of differential equations cannot be done exactly but must be done using numerical approximations that require iterative calculations. These take time. It is not uncommon for models of this nature, for relevant industrial-scale problems, to require a few hours of time on a modern computer to perform a single evaluation of the model.

In situations where we need answers at a speed comparable to the speed of the physical phenomenon, these methods often cannot be employed, and we must resort to approximate small-scale models. In many situations, such models are quite inaccurate. Situations like this are called **real time** where it is understood that a real-time model does need some time for its calculation, but this time is negligible for the application. Sometimes, real time can mean minutes and sometimes fractions of seconds; it depends on the application.

In constructing a physical, or so-called **first principles**, model, we provide many details about the construction of the situation, implying that this situation remains as we have specified. However, during the lifetime of most industrial machines, the situation changes. Materials abrade, corrode, and develop cracks. Pipes get dented and material deposits shrink the effective diameter of them. Moving parts move less efficiently or quickly over time. Objects expand with rising temperature. A first principles model cannot (usually) take such phenomena into account.

We thus have three major problems with first principles models: (1) they take a lot of effort to make and maintain, (2) they often take a long time to compute their result, and (3) they represent an idealized situation that is not the same as reality.

The desire for approximate models made by the computer as opposed to human beings is thus driven mainly by the resource requirements needed to make and to deploy first principles models. The simplest such model is a straight line.

2.5 The straight line

To illustrate many of these points, we will begin by looking at a straight-line model. Suppose we have an ideal gas in a fixed size container. As the volume and the amount of gas do not change, the ideal gas law of $PV = nRT$ is simple linear relationship between the pressure and the temperature. We could write this in the usual form of a straight line as

$$P = mT + b$$

where the slope of the line is $m = nR/V$ and the y-intercept $b = 0$. Having knowledge of the ideal gas law means we know the answer. Suppose for a moment, that we did not know the answer. We are simply provided with empirical measurements of pressure and

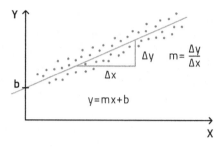

Fig. 2.9 A collection of temperature and pressure measurement plotted and found to look like a *straight line*.

temperature. As there are several of each, we will index them with the subscript i and so we have a sequence of pressures P_i and a sequence of temperatures T_i. This is illustrated in Fig. 2.9.

Based on a visual inspection, we believe that a straight-line model is a reasonable functional form for the model. That means that we must find the best slope and y-intercept for the model so that the model fits the data in the best possible way.

Before we can do that however, we must decide what "the best possible way" means, exactly. There is widespread agreement that "best" is the lowest sum of squared deviations. This is known as the **least-squares method**. In other words, we will take the squared difference between model and measurement for each observation and add them all up. That slope and y-intercept are best that makes this sum the lowest possible.

$$\min_{m,b} (P_i - mT_i - b)^2$$

In this simple case, we can solve this minimization problem explicitly (and we will omit this derivation) to obtain the answer,

$$m = \frac{\left(\sum\limits_{i=1}^{N} P_i T_i\right) - N \overline{T} \overline{P}}{\sum\limits_{i=1}^{N} T_i^2 - N\overline{T}^2}, b = \frac{\overline{P}\left(\sum\limits_{i=1}^{N} T_i^2\right) - \overline{T}\left(\sum\limits_{i=1}^{N} P_i T_i\right)}{\sum\limits_{i=1}^{N} T_i^2 - N\overline{T}^2}$$

where the overbar means taking the average over all values and N is the number of points available. We can thus calculate the best possible values of m and b directly from the empirical data. The way of proceeding from the stated minimization problem to the explicit answer for the model coefficients is the learning or **training algorithm**. It is very rare that the model is simple enough for a theoretical solution like this. In most cases, the minimization must be carried out numerically in each case. These methods are complex, time-consuming calculations that may sometimes end up in suboptimal solutions. Therefore the bulk of the machine learning literature focuses on the training algorithms.

For the time being, we note that we had to make two important decisions ourselves before executing the learning. We had to decide to use a straight-line model and we had to decide on the least-squares manner of looking for the best set of model coefficients. Those decisions are outside the realm of machine learning as they are made by a human expert.

The empirical data (T_i, P_i) had to be carefully collected. It is understood that if the experimenter made a mistake in the experiment, the value would be deleted from the dataset. Mistakes can occur in any number of forms and if they are in the dataset, then the model coefficients will contain that information, producing a somewhat false model. It is essential to clean the dataset prior to learning, so that we learn only legitimate behavior.

Finally, we must recognize that every measurement made in the real world has a measurement error. Every instrument has a measurement tolerance. The place where the instrument is put may not be the ideal place to take the measurement. Sensors drift away from their calibration, get damaged, or get covered in dirt. Environmental factors may falsify results such as the sun shining on a temperature sensor and thus temporarily inflating the measured temperature.

The dataset is therefore not just (T_i, P_i) but rather $(T_i \pm \Delta T, P_i \pm \Delta P)$. In turn, this measurement uncertainty leads to an uncertainty in the calculated model coefficients and therefore in the model itself, see Fig. 2.10. Any calculation based on an uncertain value produces an uncertain value in return. If we expect models

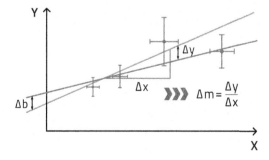

Fig. 2.10 When each data point has a measurement uncertainty in both axes (indicated by the *cross*), there is an implied uncertainty in the slope and intercept of the best *straight-line* fit.

to deliver numerical answers to our questions, it is crucial that we know how accurate we can expect those answers to be. This depends directly on the quality of the data used to make the model and the data used to evaluate the model whenever it is used. More on linear models can be found in Hastie et al. (2008).

2.6 Representation and significance

Before we analyze data, we must select or obtain the data to be analyzed. This must be done carefully with the desired outcome in mind. Suppose that you want to be able to tell the difference between circles and squares. What you would do is to collect a set of images of circles and squares and feed them into some modeling tool with the information of which images are circles and which images are squares. The modeling tool is then supposed to come up with a model that can tell the difference with some high accuracy. Suppose you give it the dataset in Fig. 2.11. That dataset contains 37 circles and 1 square.

The model that *always* says "it is a circle" would be correct in 37 out of 38 cases, using this dataset. That is an accuracy of 97%. If you went to a board meeting reporting that you have a model with a 97% accuracy for some industrial effect, you would most likely get congratulated on a job well done and the model might go live. In this case, the accuracy is not real because the dataset is not representative of the problem.

A dataset is **representative** of the problem if the dataset contains a similar number of examples of every differentiating factor of the problem. In our case that would mean a similar number of examples of circles and squares. Please note carefully that being representative of the *problem* is not the same as being representative of the *situation*. In the case of industrial maintenance, for

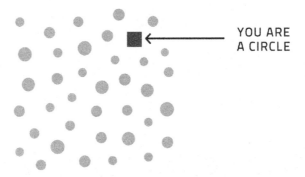

Fig. 2.11 Sample dataset for distinguishing between *circles* and *squares*. It is composed of *37 circles* and *1 square*.

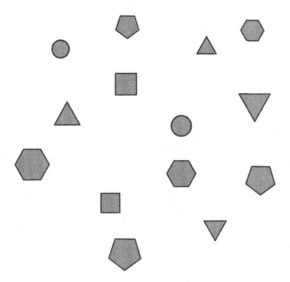

Fig. 2.12 Sample dataset for distinguishing between *circles* and *squares* containing other shapes that make the problem harder.

example, we may want to distinguish various failure modes of equipment. That is the problem. However, most of the time, there is no failure mode at all because the equipment is functioning normally. That is the situation. In collecting a dataset representative of the failure modes therefore, we must de-emphasize the normal condition even though it is the typical condition.

Now consider the same task of distinguishing between circles and squares but with the dataset given in Fig. 2.12. In addition to some circles and squares, we now also have some other shapes. This is just distracting because these are not part of the problem statement. This is not just useless data. It is going to cause harm to the model because the model will try to incorporate this into the model and thus distort the internal representation of circles and squares. These disturbance variables should be excluded because they are not significant for the problem.

A dataset is **significant** for the problem if it contains only those factors that are important for the problem. In this case, this means excluding all other shapes from the dataset.

In industrial applications, the dataset should be both representative and significant for the problem at hand. This means that we must gather some domain knowledge to ask which tags are important for the effect that we want to model, either directly or indirectly. This will make the dataset more significant. We must also ask when certain effects were observed so that we can include and exclude parts of the entire timeline into our dataset to make it more representative of the problem. For instance, we will usually want to exclude any downtime of the facility as this is usually not representative of any conditions we want to model. Conditions that are not at steady state may not be wanted either. These decisions must however be taken with the desired outcome in mind.

2.7 Outlier detection

An empirical dataset usually includes points that we do not want to show to a machine learning algorithm. These are atypical situations or the result of faulty sensors. All such points are called outliers, i.e., they lie outside of the phenomena we want to consider. They must be removed from the dataset. Of course, we can do that with domain knowledge, but this takes effort. There are automated methods to recognize outliers.

One way to do this is to cluster the data in an unsupervised manner. One of the most popular methods is the **k-means clustering** method. We must specify the number of clusters that we want to find a priori and that is the major liability in this method. It creates that many points in the dataspace, called the centers, and associates points to each one by distance. We judge the goodness of a cluster by measuring the internal similarity and the difference between different clusters. Based on this, the centers are moved around until the method converges. Individual points that do not fit easily into this scheme are outliers, see Fig. 2.13. For more on data preparation, see Aggarwal (2015).

2.8 Residuals and statistical distributions

When we have a model $\hat{y}_i = f(\underline{x}_i)$ relative to some dataset $(\underline{x}_i, y_i)$, we ask how good it is. There are many ways to quantify an answer to this question. Remember that we created the model by choosing model coefficients so that the squared difference between measurement and model is a minimum. Because of this, it makes sense to judge the quality of this model versus another model on the grounds of **root-mean-square-error or *RMSE***,

$$RMSE = \sqrt{\frac{\sum_{i=1}^{N} (\hat{y}_i - y_i)^2}{N}}$$

This measure takes its name from first computing the error e_i, also known as the **residual**, between the measurement y_i and the

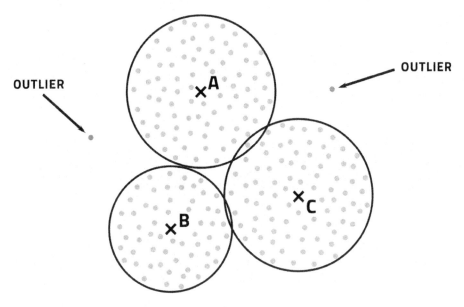

Fig. 2.13 One method to detect outliers by creating clusters.

model output $\hat{y}_i$, i.e., $e_i = \hat{y}_i - y_i$. The error is then squared so that we do not distinguish between deviations above and below the measurement. We then average these values over the whole data-set and take the square root so that the final answer is in the same units of measurement as the original value.

A related measure is the **mean-absolute-error** or *MAE*,

$$MAE = \frac{\sum\limits_{i=1}^{N} |\hat{y}_i - y_i|}{N}$$

which is often numerically quite similar to the *RMSE* but can be considered a slightly more robust measure.

As the model and measurement are supposed to be equal in an ideal world, it makes sense to compute the linear **correlation coefficient** or R^2 between them, which should be very close to 1 for a good model.

$$R^2 = \frac{\sum\limits_{i=1}^{N} (y_i - \bar{y})(\hat{y}_i - \bar{\hat{y}})}{\sqrt{\sum\limits_{i=1}^{N} (y_i - \bar{y})^2} \sqrt{\sum\limits_{i=1}^{N} (\hat{y}_i - \bar{\hat{y}})^2}}$$

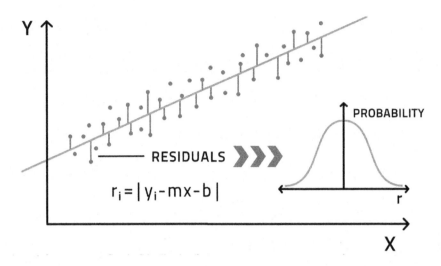

Fig. 2.14 Residuals are the differences between the data and the model.

Apart from these numerical measures of goodness, it is best to examine the **probability distribution function** of the residuals e_i, see Fig. 2.14 for an illustration. The residuals are the vertical lines between the data points and the model that has been fitted to the data. A residual can be a positive number if the data point is above the model and a negative number of the data point is below the model.

To analyze the residuals, find the smallest and largest residual. Break up the range between smallest and largest residual into several bins; let's say 100 bins which is usually sufficient for practical purposes. Go through all residuals and put them into the appropriate bin. So, each bin has a count for the number of residuals that fell into this bin. Divide the number in each bin by the total number of residuals. The result is a histogram that has been normalized to a total sum equal to 1. As the number of bins get larger and larger, this histogram becomes the theoretical probability distribution function of residuals, see Fig. 2.15. For practical purposes, the histogram is fine.

The best possible distribution that we can observe in the distribution of residuals is called the **normal distribution or Gaussian distribution**, see Fig. 2.15. This distribution has several important characteristics:

1. It is centered around zero, i.e., the most common residual is vanishingly small.
2. Its width is small, i.e., the residuals are all quite close to the (small) mean value.
3. It is symmetric, i.e., there is no systematic difference in the deviations of the model above and below the data.

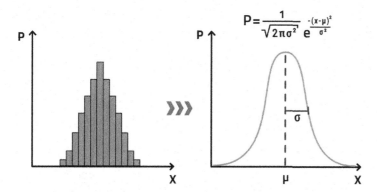

Fig. 2.15 As the number of bins in the histogram (left) gets large, it becomes the probability distribution function. The best distribution is the so-called normal distribution (right).

4. It falls off exponentially on either side of the mean so that the probability of finding a residual far from the mean is very small. This also implies that there should only be one peak in the distribution. If the distribution looks like a mountain range instead of a solitary peak, there is a fundamental problem with the model.

While the distribution can be drawn and looked at graphically, these four important characteristics can also be quantified in numerical terms and they are known as the first four moments of the distribution. In the following definitions, recall that the residual error is $e_i = \hat{y}_i - y_i$.

The first moment μ_0 is the **mean** of the distribution. This should be very close to zero.

$$\mu_0 = \bar{e} = \frac{1}{N} \sum_{i=1}^{N} e_i$$

The second moment μ_1 is the **variance** of the distribution. Most commonly, we look at the **standard deviation** $\sigma = \sqrt{\mu_1}$ that is the square root of the variance. This is easier to interpret because it has the same units of measurement as the model. The standard deviation is a measure for the width of the distribution. Ideally this value is comparable to the measurement uncertainty of the empirical data. If this is much larger than that, the model should be reassessed.

$$\mu_1 = \frac{1}{N} \sum_{i=1}^{N} (e_i - \bar{e})^2$$

The third moment μ_2 is the **skewness** of the distribution. If the distribution is symmetric, this value will be zero. If the left side of the distribution contains more data points, the skewness will be negative, and vice versa. Ideally, this number is very small.

$$\mu_2 = \frac{\frac{1}{N} \sum_{i=1}^{N} (e_i - \bar{e})^3}{\left(\frac{1}{N-1} \sum_{i=1}^{N} (e_i - \bar{e})^2 \right)^{3/2}}$$

The fourth moment μ_3 is the **kurtosis** of the distribution. The kurtosis of the normal distribution is 3 and so what we define as follows is known as the **excess kurtosis**. If your distribution has an excess kurtosis close to zero, it is like the normal distribution and that is a good thing. If it has a positive excess kurtosis, then there are more data points in the tails than the normal distribution has. This is a common observation in industrial practice. If it has negative excess kurtosis, then there are fewer data points in the tails than the normal distribution has.

$$\mu_3 = \frac{\frac{1}{N} \sum_{i=1}^{N} (e_i - \bar{e})^4}{\left(\frac{1}{N} \sum_{i=1}^{N} (e_i - \bar{e})^2 \right)^2} - 3$$

Because of outliers and the fact that exceptional circumstances are not quite as rare as we would hope, real datasets typically have heavier tails (positive excess kurtosis) than the normal distribution. This is the central burden on the model to overcome. If you get a model with low kurtosis, then the atypical situations are well taken care of.

2.9 Feature engineering

The basic task of modeling is to find some function that will take in a vector of tags and compute what we want to know. The independent variables are assumed to be sensor measurements in our industrial context, i.e., they are numbers resulting from some sort of direct measurement; they are tags. Sometimes these sources of information are called **features** of the dataset (Guyon and Elisseeff, 2003).

In many cases, we possess some domain knowledge that allows us to create further inputs to the function that are not new variables but rather derived variables. For example, a distillation column generally has a pressure measurement at the bottom and the top of the column. The reason is that the differential pressure over the column is an interesting variable. This is a piece of domain knowledge that we have. The differential pressure is itself not a measurement, but a calculation based on two pressure measurements. It is an additional feature of the dataset that we create based on our understanding of the situation. Practically, we create a new column in our dataset equal to the difference in the two pressures. This may even be a tag in the control system.

We could provide the modeling procedure with both pressure measurements and expect it to learn that it is their difference that has an important impact on the outcome. Otherwise, we could also create a derived variable that is equal to this difference and feed it into the modeling procedure ourselves. The difference is twofold.

First, providing raw data only, means that the modeling procedure must figure out basic variable transformations and this requires information. Information is contained in the dataset, but it is finite. As some information is used up to learn these transformations, less information is available to learn other dependencies meaning that the final model will most likely perform less well.

Second, these basic transformations must be represented in the model and this requires parameters. For a certain limited number of parameters in the model, this restricts the expressiveness of the model unnecessarily.

It is therefore to our advantage to include in the dataset all derived quantities that are known to be of interest and this requires domain knowledge and some thought. It is well worth doing because this creates better and more robust models. This process is known as **feature engineering**. The amount of effort that goes into this must not be underestimated and should be regarded as an essential step in the treatment of industrial datasets.

The essential point to understand is that the information present in a dataset is limited and that there is a **principle of diminishing returns** in the size of the dataset. When the dataset is small, each new data point adds information. At some point, the dataset is rich enough that new data points do not add information. The dataset is saturated in information. Despite being able to grow the dataset with effort and money, you cannot provide new information. Recognizing that information is a limited resource, we must be careful how we spend it. Spending it on learning to establish features already known to human experts is wasteful.

Having looked at adding some features, it is also necessary and instructive to remove some features as they do not necessarily encapsulate any information.

2.10 Principal component analysis

Suppose that we have a dataset with three independent variables as displayed in Fig. 2.16. We can see that this dataset has three rough clusters of data points in it. As this is the main visible structure in the dataset, we want to retain this, but we ask if we are able to lower the dimensionality of the dataset somewhat.

If we project the dataset onto a single one of the three coordinate axes, we get the result displayed in Fig. 2.17 on the top right, i.e., all three such projections are just blobs of data in which the clustered structure is lost.

However, if we were to first rotate the dataset in three-dimensional space to align it with the axes indicated by the two arrows in Fig. 2.16 and then project them onto the axes, we get the result displayed in Fig. 2.17 on the bottom right. The rotated dataset projected onto the two-dimensional plane defined by the two axes is displayed in Fig. 2.17 on the left. We easily see that the two-dimensional representation captures most of the information and the one-dimensional representation indicated by "pc1" also retains the clustering structure. We are thus able to reduce the dimensionality of the dataset to either two or even one dimension by rotating it appropriately.

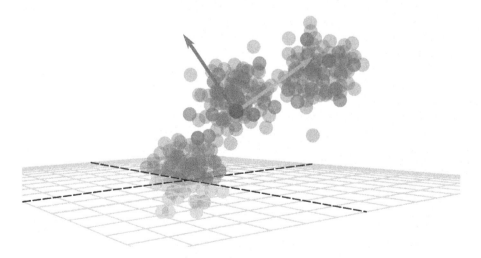

Fig. 2.16 A three-dimensional dataset to illustrate principal component analysis.

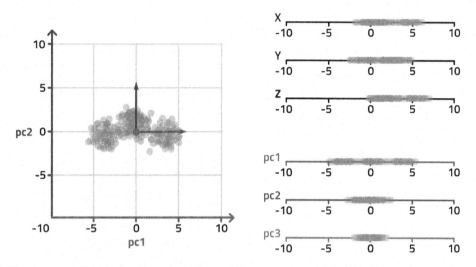

Fig. 2.17 The dataset of Fig. 2.16 is projected onto its coordinate axes (top right) and its principal components (bottom right) as well as the plane of the first two principal components (left).

The rotation we are looking for is such that the new axes are along the directions of greatest variance in the dataset. This is displayed easily in Fig. 2.18 where a dataset that we can fit with a straight line is displayed. The direction of that straight-line model is the direction of greatest variance in the dataset and could be used to represent most of the information. The perpendicular direction then represents the direction of second greatest variance. These directions in order from the most significant to the least significant in the sense of variance are called the principal components and the method that computes them is called **principal component analysis** or **PCA**. A discussion of the

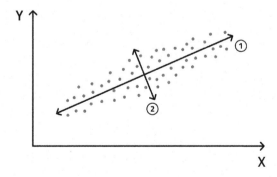

Fig. 2.18 An indication of the principal components of a simple linear dataset.

calculations involved is beyond this book but many software packages include this procedure as an essential data science feature and so it can be readily carried out in practice.

The variances can be quantified precisely, and they can be added up. We can then ask: How many principal components must we keep in order to explain a certain amount, 99% say, of the total variance in the original dataset? With industrial datasets where many of the variables are related to each other in various ways, we often find that the dimensionality can be reduced by as much as one-half, without significantly reducing the information content of the dataset. The reason for this is simple: As the tags originate from a single coordinated system, they duplicate a lot of information. More information can be found in James et al. (2015).

Dimensionality reduction is an important preprocessing step for machine learning and data science in general. It reduces the complexity of the dataset and the model, which generally leads to more robust models. Principal component analysis is recommended in virtually all modeling situations.

2.11 Practical advices

When faced with a data analysis challenge, what should we do in practice?

First, define very clearly what the desired outcome is. Do you need to forecast something at a future time? Do you need to characterize the state of something into various categories? Do you need decision support in ranking options? Do you want a graphical display of various pieces of information? How accurate does this analysis have to be in order to achieve its purpose? How often does the analysis need to get done—once a day or in real time?

Second, decide which data sources you need to tap into to get the data for this outcome. This is going to be the control system or historian and so the real question is which tags do you want to look at? It will almost invariably be a small selection of all available tags and so you need to make a careful choice based on domain knowledge. Tools like correlation analysis can help in this selection. Having got this far, you can extract historical data from your database.

Third, decide how you are going to measure success. Such metrics are available from a data analytics point of view as discussed in this chapter, but success metrics are also of a practical nature such as (1) how much money is this earning, (2) how many predictions were correct, (3) how many proposals made by the analysis were adopted, (4) how many of the people who need to use/implement/follow this system accept it willingly.

Fourth, this dataset is now raw and must be cleaned before analysis can take place. Data that is irrelevant should be removed, such as data when the plant was offline or being maintained. If a sensor was temporarily broken, those data points might need to be interpolated or removed as well. Unusual operating conditions may need to be removed, depending on the use case. Most data will be quite similar, and some data will represent rare conditions. For the dataset to be representative and significant relative to the use case, you may need to resample the dataset so that analysis can pick up the dynamics you want to focus on.

Fifth, based on the clean dataset, think if there are some combinations of these data that would provide useful information to the analysis. If you know, for instance, that the ratio of two quantities is crucial for the outcome, then you should introduce a further column in the data matrix with that ratio. Material balance may be an important aspect and so calculating material in and material out are examples of important features that can guide the analysis.

Sixth, analysis will benefit from dealing with as few data columns as possible that nonetheless contain all the information needed. Dimensionality reduction methods such as principal component analysis can do this elegantly.

At this point, you have a clean, streamlined dataset that includes all the information and domain knowledge you can give it. You are now ready to perform the analysis of this data relative to your use case and you know how to assess the success of this analysis so that if you try more than one method, you can compare them.

Please note that we have not talked about machine learning yet. In practice, getting to this point requires 80% of the person-hours for the whole project. Getting to a good dataset is the main work effort. Recognizing this alone will increase your chances of success greatly. Let's talk about machine learning next.

References

Aggarwal, C.C., 2015. Data Mining. Springer, Heidelberg.

Guyon, I., Elisseeff, A., 2003. An introduction to variable and feature selection. J. Mach. Learn. Res. 3, 1157–1182.

Hamilton, J.D., 1994. Time Series Analysis. Princeton University Press, Princeton.

Hastie, R., Tibshirani, R., Friedman, J., 2008. The Elements of Statistical Learning, second ed. Springer, Heidelberg.

James, G., Witten, D., Hastie, T., Tibshirani, R., 2015. An Introduction to Statistical Learning. Springer, Heidelberg.

Press, W.H., Teukolsky, S.A., Vetterling, W.T., Flannery, B.P., 2007. Numerical Recipes, third ed. Cambridge University Press, Cambridge.

Pyle, D., 1999. Data Preparation for Data Mining. Morgan Kaufmann, San Francisco.

3

Machine learning

Patrick Bangert[a,b]

[a]Artificial Intelligence, Samsung SDSA, San Jose, CA, United States
[b]Algorithmica Technologies GmbH, Bad Nauheim, Germany

In this chapter, we give an introduction to machine learning, its pivotal ideas, and practical methods as applied to industrial datasets. The dataset was discussed in the previous chapter and so here we assume that a clean dataset already exists alongside a clear problem description. Some principal ideas are introduced first, after which we discuss some different model types, and then discuss how to assess the quality and fitness for purpose of the model.

Machine Learning and Data Science in the Power Generation Industry. https://doi.org/10.1016/B978-0-12-819742-4.00003-2
© 2021 Elsevier Inc. All rights reserved.

3.1 Basic ideas of machine learning

Machine learning is the name given to a large collection of diverse methods that aim to produce models given enough empirical data only. They do not require the use of physical laws or the specification of machine characteristics. They determine the dependency of the variables among each other by using the data, and only the data.

That is not to say that there is no more need for a human expert. The human expert is essential but the way the expertise is supplied is very different to the first principles model. In machine learning, domain expertise is supplied principally in these four ways:

1. Providing all relevant variables and excluding all irrelevant variables. This may include some elementary processed variables via feature engineering, e.g., when we know that the ration between two variables is very relevant, we may want to explicitly supply that ratio as a column in the data table.
2. Providing empirical data that is significant and representative of the situation.
3. Assessing the results of candidate models to make sure that they output what is expected.
4. Explicitly adding any essential restrictions that must be obeyed.

These inputs are important, but they are easily supplied by an expert who knows the situation well and who knows a little about the demands of data science.

The subject of **machine learning** has three main parts. First, it consists of many prototypical models that could be applied to the data at hand. These are known by names such as neural networks, decision trees, or k-means clustering. Second, each of these comes with several prescriptions, so-called **algorithms** that tell us how to calculate the model coefficients from a dataset. This calculation is also called **training** the model. After training, the initial prototype has been turned into a model for the specific dataset that we provided. Third, the finished model must be deployed so that it can be used. It is generally far easier and quicker to evaluate a model than to train a model. In fact, this is one of the primary features of machine learning that make it so attractive: Once trained, the model can be used in real time. However, it needs to be embedded in the right infrastructure to unfold its potential.

Associated to machine learning are the two essential topics that are at the heart of **data science**. First, the data must be suitably prepared for learning, which we discussed in Chapter 2. Second, the resultant model must be adequately tested, and its performance must be demonstrated using rigorous mathematical means. This preprocessing and postprocessing before and after machine learning is applied to round out the scientific part of a data science project. In addition to these scientific parts, there are managerial and organizational parts that concern collecting the data and dealing with the stakeholders of the application.

In this book, we will treat all these topics, except the training algorithms. These form the bulk of the literature on machine learning and are technically challenging. This book aims to provide an overview to the industrial practitioner and not a university course on machine learning. The practitioner has access to various computer programs that include these methods and so a detailed understanding of how a model is trained is not necessary. On the other hand, it is essential that a practitioner understand what must be put into such a training exercise and how to determine whether the result is any good. So, this will be our focus.

Methods of machine learning are often divided into two groups based on two different attributes. One grouping is into **supervised** and **unsupervised** methods. The other grouping is into **classification** and **regression** methods.

Supervised methods deal with datasets for which we possess empirical data for the inputs to the model as well as the desired outputs of the model. Unsupervised methods deal with datasets for which we only have the inputs. Imagine teaching a child to add two numbers together. Supervised learning consists of examples like $1+2=3$ and $5+4=9$ whereas unsupervised learning consists of examples like $1+2$ and $5+4$. It is clear from this difference that we can expect supervised methods to be much more accurate in reproducing the outcome. Unsupervised methods are expected to learn the structure of the input data and recognize some patterns in them.

Fig. 3.1 illustrates the difference in the context of learning the difference between two collections of data points. In the first example, we know that circles and triangles are different and must learn where to divide the dataspace between them. In the second example, we only have points and must learn that they can be meaningfully divided into two clusters such that points in one cluster are maximally similar and points in different clusters are maximally different from each other—for some measure of similarity that makes sense in the context of the problem domain.

The difference between **classification** and **regression** methods is similarly fundamental. Classification methods aim to place a data point into one of several groups while regression methods aim to compute some numerical value. Fig. 3.2 illustrates the difference. In the first example, we only wish to draw a dividing line

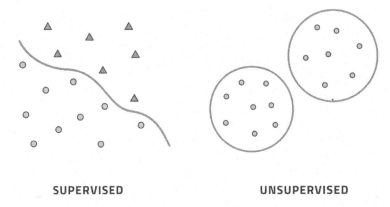

SUPERVISED   UNSUPERVISED

Fig. 3.1 An illustration of the difference between supervised and unsupervised learning.

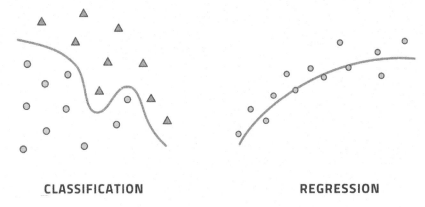

CLASSIFICATION   REGRESSION

Fig. 3.2 An illustration of the difference between classification and regression methods.

between the circles and the triangles, differentiating one category from the other. In the second example, we want to reproduce a continuous numeric value as accurately as possible.

This distinction is closely related to the difference between a **categorical** variable and a **continuous** variable. A categorical variable is equal to one of several values, like 1, 2, or 3, where the values indicate membership in some group. The difference between two values is therefore meaningless. If data point A has value 3 and data point B has value 2, all this means is that A belongs to group 3 and B belongs to group 2. Taking the difference of the values, $3-2=1$, has no meaning, i.e., the difference between the two data points does not necessarily belong to group 1 nor does group membership necessarily even make sense for the difference.

A continuous variable is different in that it measures a quantity and taking the difference does mean something real. For instance, if data point A has a flowrate of 1 ton per hour and data point B has a flowrate of 2 tons per hour, we can take the difference $2 - 1 = 1$ and the difference is 1 ton per hour of flow, which is a meaningful quantity that we can understand.

Machine learning has many methods and all methods are either supervised or unsupervised, and either classification or regression. Any task that uses machine learning can also be divided into these categories. These two groupings are the first point of departure in selecting the right method to solve the problem at hand. So, ask yourself

1. Do you have data for the result you want to compute, or only for the factors that go into the computation?
2. Is the result a membership in some category, or a numerical value with intrinsic meaning?

We will present methods for all these possibilities guided by the state-of-the-art and industrial experience. This book will not present an exhaustive list of all possible methods as this would go well beyond our scope.

As we need to be brief on the technical aspects of machine learning, here are some great books for further reading on machine learning in and of itself. A great book on the ideas of machine learning without diving into its technical depths is Domingos (2015). If you are more interested in the general economic ramifications of the field, an excellent presentation is in Agrawal et al. (2018). More mathematical books that present a great overview are Mitchell (2018), Bishop (2006), MacKay (2003), and Goodfellow et al. (2016).

3.2 Bias-variance-complexity trade-off

One of the most important aspects in machine learning concerns the quality of the model that you can expect, given the quality of the data you have to make the model. Three aspects of the model are important here.

The **bias** of the model is an assessment of how far off the average output of the model is from the average expected output. The **variance** of the model is the extent to which inputs that are very close to each other result in outputs that are far from each other. The **complexity** of the model is generally measured by the number of parameters that must be determined using a machine learning algorithm from the empirical data provided for training.

The concepts of bias and variance are depicted in Fig. 3.3. Imagine throwing darts at a target. If the darts are all close together,

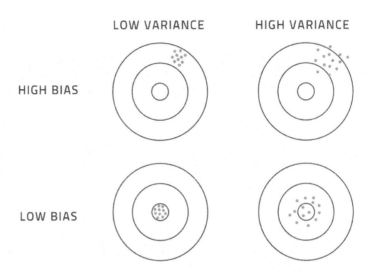

Fig. 3.3 The concepts of bias and variance illustrated by throwing darts at a target.

the variance is low. If the darts are, on average, close to the center of the target, the bias is low. Of course, we want a model with low bias and low variance.

In order to get such a model, we first choose a type of model, i.e., a formula with unknown parameters for which we believe that it can capture the full dynamics of our dataset if only the right values for the parameters are found. An example of this is the straight-line model that naturally has two parameters for one independent variable. We shall meet several other models later on that are more expressive because they have more parameters. Having thought about the straight line, we can easily understand that a quadratic polynomial (three parameters for one independent variable) can model a more complex phenomenon than a straight line.

Choosing a model that has very few parameters is not good because it will not be able to express the complexity present in the dataset. The model performance in bias and variance will be poor because the model is too simple. This is known as **underfitting**.

However, choosing a very complex model with a great many parameters is not necessarily the solution because it may get so expressive that it can essentially memorize the dataset without learning its structure. This is an interesting distinction. Learning something suggests that we have understood some underlying mechanism that allows us to do well on all the examples we saw while learning, *and* any other similar tasks that we have *not* seen while learning. Memorization will not do this for us and so

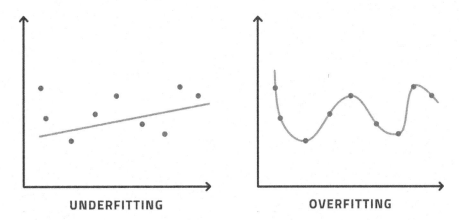

UNDERFITTING   OVERFITTING

Fig. 3.4 If the model is too simple, we will underfit. If the model is very complex, it will overfit.

it is generally thought of as a failed attempt to model something. It is known as **overfitting**. See Fig. 3.4 for an illustration of underfitting and overfitting.

There must exist some optimal number of parameters that is able to capture the underlying dynamics without being able to simply memorize the dataset. This is some medium number of parameters.

If the model reproduces the data used in its construction poorly, then the model is definitely bad, and this is usually either underfitting or a lack of important data. The fix is to either use a more complex model, to get more data points, or to search for additional independent variables (additional sources of information). The data used in the construction of the model is the **training data** and the difference between the model output and the expected output for the training data is the **training error**.

To properly assess model performance, we cannot use the training data, however. We must use a second dataset that was not used to construct the model. This is the **testing data** that results in a **testing error**. The testing data is often known as **validation data**. More complex training algorithms need two datasets, one to train on and another to assess when to stop training. In situations like this, we may need to create three datasets and these are then usually called training, testing, and validation datasets where the testing dataset is used to assess whether training is done and the validation dataset is used to assess the quality of the model.

Fig. 3.5 displays the typical model performance as a function of the model complexity. As the model gets more complex, the training error decreases. The testing error decreases at first and then

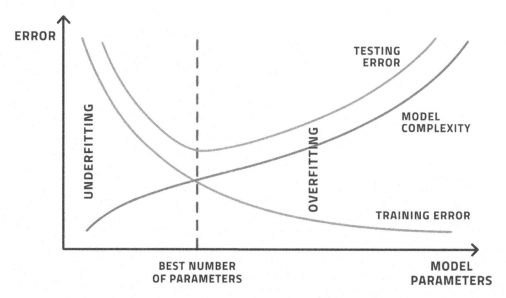

Fig. 3.5 As the model gets more complex, the training and testing errors change, and this indicates an optimal number of parameters.

increases again as the model starts to overfit. There is a point, at minimum testing error, when the model achieves its best performance. This is the performance as measured by the bias, i.e., the deviation between computed and expected values. This has not yet taken the variance into account. Often, we find that to get better variance, we must sacrifice some bias and vice versa. This is the nature of the bias-variance-complexity trade-off.

While this picture is typical, we usually cannot draw it in practice as each point on this image represents a full training of the model. As training is usually not a deterministic process, the training must be repeated several times for each sample complexity to get a representative answer. This is also needed to measure the variance. It is, in most practical situations, an investment of time and effort no one is prepared to make.

A similar diagram, see Fig. 3.6, can be drawn in terms of variance. There are two major sources of uncertainty in machine learning. First, the uncertainty inherent in the dataset and the manner in which it was obtained. This includes both the method of measurement and the managerial process of selecting which variables to include in the first place. Second, the uncertainty contained in the model and the basic ability of it to express the underlying dynamics of the data. There is a similar compromise that must happen.

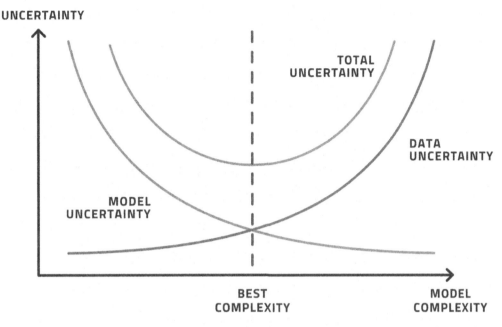

Fig. 3.6 As model complexity rises, model uncertainty decreases but data uncertainty increases. There is an optimal point in the middle.

In the end, we must choose the right model type and the right number of parameters for this model to achieve a reasonable compromise between bias and variance, avoiding both underfitting and overfitting. To prove to ourselves and other that all this has been done, we need to examine the performance on both training and testing data.

3.3 Model types

There are many types of model that we can choose from. This section will explore some popular options briefly with some hints as to when it makes sense to use them. To learn more about diverse model types and machine learning algorithms, we refer to Goodfellow et al. (2016).

3.3.1 Deep neural network

The **neural network** is perhaps the most famous and most used technique in the arsenal of machine learning (Hagan et al., 1996). The recent advent of **deep learning** has given new color

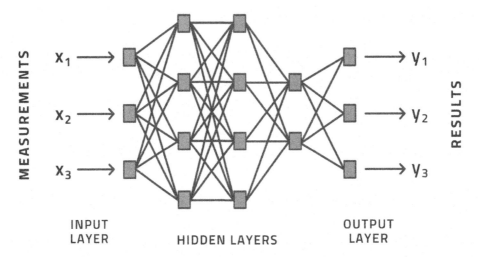

MEASUREMENTS

$x_1 \longrightarrow$

$x_2 \longrightarrow$

$x_3 \longrightarrow$

$\longrightarrow y_1$

$\longrightarrow y_2$

$\longrightarrow y_3$

RESULTS

INPUT
LAYER

HIDDEN LAYERS

OUTPUT
LAYER

Fig. 3.7 Schematic illustration of a deep neural network.

to this model type through novel methods of training its parameters. Training methods are beyond the scope of this book, however.

While it took its inspiration from the network of neurons in the human brain, the neural network is merely a mathematical device. A schematic diagram like the one shown in Fig. 3.7 is commonly displayed for neural networks. On the left side, the input vector x enters the network. This is called the **input layer** and has several neurons equal to the number of elements in the input vector. On the right side, the output vector y exits from the network. This is called the **output layer** and has several neurons equal to the number of elements in the output vector; this is the result of the calculation. In between them are several **hidden layers**. The number of hidden layers and the number of neurons in each are up to us to choose and is referred to as the **topology** of the neural network.

In bringing the data from one layer to the next, we multiply it by a matrix, add a vector, and apply a so-called **activation function** to it. These are the parameters for each layer that the machine learning algorithm must choose from the data so that the neural network fits to the data in the best possible way. A neural network with two hidden layers, therefore, looks like

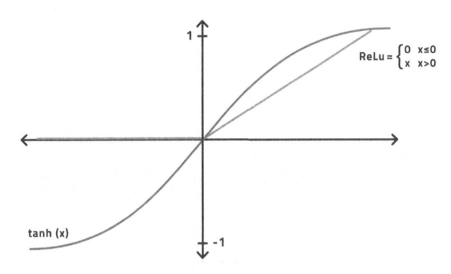

$$\text{ReLu} = \begin{cases} 0 & x \le 0 \\ x & x > 0 \end{cases}$$

tanh (x)

Fig. 3.8 Two popular activation functions for a neural network.

$$\underline{y} = \underline{A}_2 \cdot f(\underline{A}_1 \cdot f(\underline{A}_0 \cdot \underline{x} + \underline{b}_0) + \underline{b}_1) + \underline{b}_2$$

The activation function $f(\cdots)$ is not learnt but decided on beforehand. Considerable research has been put into this choice. While older literature prefers $\tanh(\cdots)$, we now have evidence that *ReLu* $(\cdots)$ performs significantly better in virtually all tasks. They are both displayed in Fig. 3.8.

$$ReLu(x) = \begin{cases} 0 & x \le 0 \\ x & x > 0 \end{cases}$$

The neural network is not some magical entity. It is the earlier equation. After we choose values for the parameter matrices $\underline{A}_i$ and vectors $\underline{b}_i$, the neural network becomes a specific model. The main reason it is so famous is that it has the property of **universal approximation**. This property states that the formula can represent any continuous function with arbitrary accuracy. There are other formulae that have this property, but neural networks were the first ones that were practically used with this feature in mind. In practice, this means that a neural network can represent your data. What the property does not do is tell us how many layers and how many neurons per layer to choose. It also does not tell us how to find the correct parameter values so that the formula does in fact represent the data. All we know is that the neural network *can* represent the data.

Machine learning as a research field is largely about designing the right learning algorithms, i.e., how to find the best parameter values for any given dataset. There are multiple desiderata for it, not just the quality of the parameter values. We also want the training to be fast and to use little processing memory so that training does not consume too many computer resources. Also, we want it to be able to learn from as few examples as possible. Beyond the hype of "big data," there is now significant research into "small data."

Many practical applications provide a certain amount of data. Getting more data is either very difficult, expensive, or virtually impossible. In those cases, we cannot get out of machine learning difficulties by the age-old remedy of collecting more data. Rather, we must be smarter in getting the information out of the data we have. That is the challenge of small data and it is not yet solved. We mention it here to put the property of universal approximation into proper context. It is a nice property to be sure, but it has little practical value.

The neural network is the right model to use if the empirical observations x are independent of each other. For example, if we take a manual sample from flue gas once per day and perform a laboratory analysis on it, we can assume that today's observation is largely independent of yesterday's observation but may well depend on the pressure and temperature right now. In this context, we may use pressure and temperature as inputs and expect the neural network to represent the laboratory measurement.

If, however, time delays do play a role, then we must incorporate time into the model itself. Let's say we change the coal mill settings in a coal power plant and measure its burner profile, we will see that it responds to our action a few minutes later. If we measure these quantities every minute, then the observations are not independent of each other because the observation a few minutes ago causally brought about the current observation. Applying the neural network to such datasets is not a good idea because the neural network cannot represent this type of correlation. For this, we need a recurrent neural network.

3.3.2 Recurrent neural network or long short-term memory network

A **recurrent neural network** is very similar in spirit to the regular, the so-called feed-forward, neural network. Instead of having connections that move strictly from left to right in the diagram, it

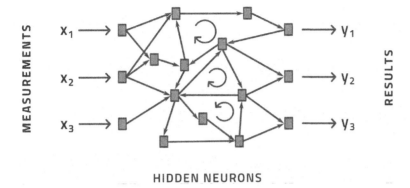

MEASUREMENTS

$x_1 \longrightarrow$ y_1

$x_2 \longrightarrow$ y_2

$x_3 \longrightarrow$ y_3

RESULTS

HIDDEN NEURONS

Fig. 3.9 Schematic illustration of a recurrent neural network.

also includes connections that amount to cycles, see Fig. 3.9. These cycles act as a memory in the formula by retaining—in some transformed way—the inputs made at prior times.

As mentioned in the previous section, the purpose of recurrence is to model the interdependence between successive observations. In practice such interdependence is virtually always due to time and represents some mechanism of causation or control. If my foot touches the brake pedal in my car, the car slows down a short time later. Modeling the dependence of car speed on the position of the brake pedal can lead to important models, i.e., the advanced process control behind the automatic braking system in your car. The time delay involved is important as the car will have moved some distance in that time and so braking must begin sufficiently early so that the car stops before it hits something. It is cases such as this for which the recurrent neural network has been developed.

There are many different forms of recurrent neural networks differing mainly in how the cycles are represented in the formula. The current state of the art is the **long short-term memory** network, or **LSTM** (Hochreiter and Schmidhuber, 1997; Yu et al., 2019; Gers, 2001). This network is made up of cells. A schematic of one cell is shown in Fig. 3.10. These cells can be stacked both horizontally and vertically to make up an arbitrarily large network of cells. Once again, it is up to the human data scientist to choose the topology of this network.

The inner working of one cell is defined by the following equations that refer to Fig. 3.10.

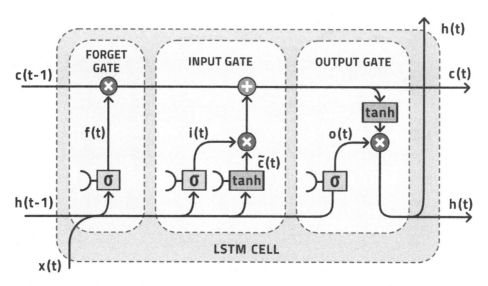

Fig. 3.10 Schematic illustration of one cell in an LSTM network.

$$\underline{f}(t) = \sigma\left(\underline{W}_{fh} \cdot \underline{h}(t-1) + \underline{W}_{fx} \cdot \underline{x}(t) + \underline{b}_f\right)$$
$$\underline{i}(t) = \sigma(\underline{W}_{ih} \cdot \underline{h}(t-1) + \underline{W}_{ix} \cdot \underline{x}(t) + \underline{b}_i)$$
$$\underline{\tilde{c}}(t) = \tanh\left(\underline{W}_{\tilde{c}}h \cdot \underline{h}(t-1) + \underline{W}_{\tilde{c}}x \cdot \underline{x}(t) + \underline{b}_{\tilde{c}}\right) \quad (t)$$
$$\underline{c}(t) = \underline{f}(t) \cdot \underline{c}(t-1) + \underline{i}(t) \cdot \underline{\tilde{c}}$$
$$\underline{o}(t) = \sigma(\underline{W}_{oh} \cdot \underline{h}(t-1) + \underline{W}_{ox} \cdot \underline{x}(t) + \underline{b}_o)$$
$$\underline{h}(t) = \underline{o}(t) \cdot \tanh \underline{c}(t)$$

The various $\underline{W}$ and $\underline{b}$ are the weight and bias parameters of the LSTM cell. All weights and biases of the entire network must be tuned by the learning algorithm. The evolution of the internal state of the network $\underline{h}(t)$ is the memory of the network that can encapsulate the time dynamics of the data.

The network is trained by providing it with a time series $\underline{x}(t)$ for many successive values of time t. The model outputs the same time series but at a later time $t+T$, where T is the forecast horizon. This is the essence of how LSTM can forecast a time series

$$\underline{x}(t+T) = \mathcal{L}_t(\underline{x}(t))$$

It is generally not a good idea to model the next time step and then to chain the model

$$\underline{x}(t+3) = \mathcal{L}_{t+2}(\underline{x}(t+2)) = \mathcal{L}_{t+2}(\mathcal{L}_{t+1}(\underline{x}(t+1)))$$
$$= \mathcal{L}_{t+2}(\mathcal{L}_{t+1}(\mathcal{L}_t(\underline{x}(t))))$$

Because this compounds errors and makes for a very unreliable model if we want to predict more than one or two time steps into the future. It is much better to choose the forecast horizon to be some larger number of time steps. Having done that, the intermediate values can always be computed using the same model with input vectors from earlier in time.

3.3.3 Support vector machines

The **support vector machine** (SVM) is a technique mainly used for classification both in a supervised and unsupervised manner and can be extended to regression as well (Cortes and Vapnik, 1995). The idea, as illustrated in Fig. 3.11, is that we have a set of points that belong to one of two categories and we draw a non-linear plane through the space dividing the data into two pieces. Everything above the plane is classified as one category and everything below the plane is classified as the other. The plane itself is an interpolated spline-like object that is based on a few selected data points—the support vectors—that must be found.

The plane chosen is that which has the maximum distance from all points possible. The points at the boundary of this space form the support vectors. In practice, the classification problem may require multiple planes for a good result. As the space of

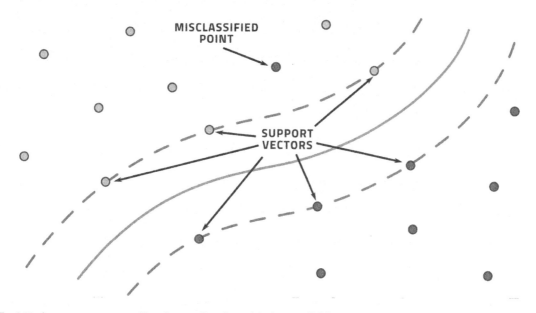

Fig. 3.11 A support vector machine draws a line through a dataset dividing one category from another as efficiently as possible.

all points gets divided into sections by a collection of planes, we may find that a few data points are isolated in one area. This is another way of identifying outliers.

3.3.4 Random forest or decision trees

The **decision tree** is familiar perhaps from management classes where it is used to structure decision making. At the base of the tree is some decision to be made. At each branching, we ask some question that is relatively easy to answer. Depending on the answer, we go down one branch as opposed to another and eventually reach a point on the tree that has no more branches, a so-called leaf node. This leaf node represents the right answer to the original decision.

Fig. 3.12 illustrates this using a toy example. The question is whether the equipment needs maintenance or not. The left-hand tree starts by analyzing the equipment age in years. If this is larger than 10 years, we go down the left-hand path and otherwise the right-hand path. The left-hand path terminates at a leaf node with the value 1. The right-hand path leads to a secondary question where we ask for the number of starts of the equipment. If this is larger than 500, we go down that left-hand path leading us to the number 1/2 and otherwise we go down the right-hand path leading to the number 1/3. The number that we end up with is just a numerical score for now.

The numerical score now needs to be interpreted, which is usually done by thresholding, i.e., if the score is greater than some fixed threshold, we decide that the equipment needs preventative maintenance and otherwise we decide that it can operate a little

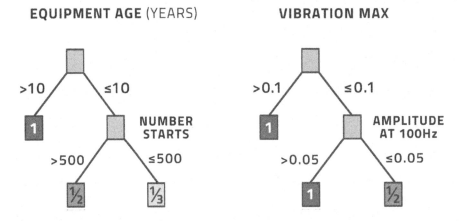

Fig. 3.12 An example of two decision trees that can be combined into a random forest.

longer without inspection. This is an example of a decision tree in the industrial context (Breiman et al., 1984).

We can combine several decision trees into a single decision-making exercise. Fig. 3.12 displays a second tree that operates in the same way. Having gone through all the trees, we add up the scores and threshold the final result. If the left tree leads to ½ and the right tree leads to 1, then the total is a score of 3/2. This combination of several decision trees is called a **random forest**.

Constructing the trees, their structures, the decision factors at each branching, and the values for each branching are the model parameters that must be learnt from data. There are sophisticated algorithms for this that have seen significant evolution in recent years making random forest one of the most successful techniques for classification problems. Incidentally, the forest is far from random, of course. The forest is carefully designed by the learning method to give the best possible classification result possible. Random forests can also be used for regression tasks, but they seem to be better suited to classification tasks.

3.3.5 Self-organizing maps

A **self-organizing map (SOM)** is an early machine learning technique that was invented as a visualization aid for the human expert (Kohonen, 2001). Roughly, it belongs to the classification kind of techniques as it groups or clusters data points by similarity. The idea is to represent the dataset with special points that are constructed by the learning method in an iterative manner. These special points are mapped on a two-dimensional grid that usually has the topology of a honeycomb. The topology is important because the learning, i.e., the moving around of the special points, occurs over a certain neighborhood of the special point under consideration.

When this process has converged, the special points are arrayed in the topology in a certain way. Any point in the dataset is associated with the special point that it is closest to in the sense of some metric function that must be defined for the problem at hand. This generates a two-dimensional distribution of the entire data. It is now usually a task for the human expert to look at this distribution and determine to what extent the cells in the topology (the honeycomb cells) are similar to each other. Frequently it turns out that several cells are very similar and can be associated with the same macroscopic phenomenon.

For example, in Fig. 3.13 we begin with a topology of 22 honeycomb cells. Having done the learning and the human examination, it is determined that there are only four macroscopically

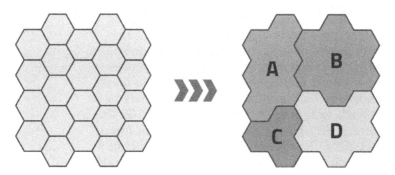

Fig. 3.13 An example of a self-organizing map.

relevant clusters present. Each cluster is represented by several cells. These cells can either have a more subtle meaning within the larger group or they can simply be a more complex representation of a single cluster than one special point, as one would do using algorithms like k-means clustering (Seiffert and Jain, 2002).

Ultimately, when one encounters a new data point, one would compute the metric distance between it and all the special points learnt. The special point closest to the new point is its representative and we know which cluster it belongs to. The SOM method has turned out to be a very good classification method in its own right. In addition to this however, the current position of a system can be plotted on the graphic and one can see the temporal evolution on the state chart, which is an elegant visual aid for any human working with the data source.

3.3.6   Bayesian network and ontology

Bayesian statistics follows a different philosophy than standard statistics and this has some technical consequences (Gelman et al., 2004). Without going into too much detail, standard statistics assesses the probability of an event by the so-called frequentist approach. We will need to collect data as to how frequently this event occurs and does not occur. Usually, this is no problem but sometimes the empirical data is hard to come by, events are rare or very costly, and so on.

Bayesian statistics concerns itself with the so-called degree of belief. This encapsulates the amount of knowledge that we have. If we manage to gather additional knowledge, then we may be able to update our degree of belief and thereby sharpen our probability estimate. The probability distribution after the update is called the posterior distribution. There is a specific rule, Bayes theorem, on how to perform this updating and there is no doubt or difficulty in its application. The tricky part is establishing the probability distribution at the very beginning, the prior distribution. Usually we assume that all events are equally likely if we have no knowledge at all or we might prime the probabilities using empirical frequencies if we do have some knowledge.

Supposing that we have gathered quite a bit of knowledge about the general situation, we can start to create multiple conditional probability distributions. These are conditional on knowing or not knowing certain information.

Let's take the case of a root-cause analysis for a mechanical defect in a gas turbine, see Fig. 3.14. We know from past experience that certain measurements provide important information relating to the root cause of this problem, e.g., the axial deviation, which is the amount of movement of the rotor away from its engineered position. This, in turn, is influenced by the bearing temperature and the maximum vibration. These two factors are influenced by other factors in turn. We can build up a network of mutually influencing events and factors. At one end of this matrix are the events we are concerned about such as the mechanical defect. At the other end are the events we can directly cause to happen such as the addition of lubricant into the system.

Based on our past experience we can establish that if the maximum vibration is above a certain limit, the chance of a rotor imbalance increases by a certain amount, and so on for all the other factors. We can then use statistics to obtain a conditional probability distribution for the mechanical defect as a function

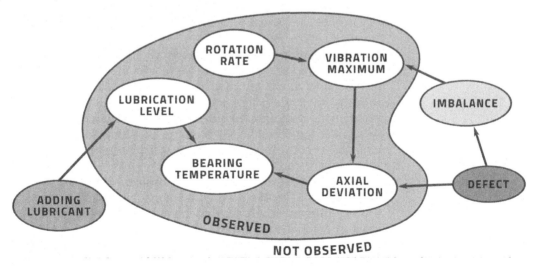

Fig. 3.14 A very simple example of a Bayesian network for deciding on the root cause of a mechanical defect in a turbine.

of all the factors. This cascading matrix of conditional probabilities is called a **Bayesian network** (Scutari and Denis, 2015). Once we have this network, it can be used like so: Initially, the chance of a mechanical defect is very small; we assess this from the frequency of observations of such defects. Then we encounter a high bearing temperature and a high vibration for a low rotation rate. Combining this information using the various conditional probabilities updates our probability of defect. If this is sufficiently high, we may release an alarm and give a specific probability for a problem. We can also compute what effect it would have on the probability if we added lubricant and, based on the result, may issue the recommendation for this specific action.

Please note that this is very different from an **expert system**. An expert system looks similar at first glance, but it is a system of rules written by human experts as opposed to the Bayesian network that is generated by an algorithm based on data. The expert system has the advantage that all components are designed by people who deeply understand the system. However, combinations of rules often have logical consequences that were not intended by those experts and these are difficult to detect and prevent. Due to this complexity, expert systems have usually failed to work, or at least be commercially relevant given all the effort that flows into them. At present, expert systems are considered an outdated technology. In comparison, Bayesian networks are constructed automatically and therefore can be updated automatically as new data becomes available. An understanding of each individual link in the chain is possible, just like an expert system, but the totality of a network can usually not be understood as a practical network is usually quite large.

Bayesian networks are great as decision support systems. How much will this factor contribute to an outcome? If we perform this activity, how much will the risk decrease? In our asset-intensive industry, the main use cases are (1) root-cause analysis for some problem; (2) deciding what to do and to avoid based on a concrete numerical target value; (3) scenario planning as a reaction to changes in consumer demand, market prices, and other events in the larger world.

A related method is called an **ontology** (Ebrahimipour and Yacout, 2015). This is not a method of machine learning but rather a way to organize human understanding of a system. We may draw a tree-like structure where a branch going from one node to another has a specific meaning such as "is a." For example, the root node could be "rotating equipment." The node "turbine" then has a link to the root node because a turbine is an example of a rotating equipment. The nodes "gas turbine," "steam turbine," or "wind turbine" are all examples of a turbine. This tree can be extended both by adding more nodes as well as adding other types of branches such as "has a" or "is a part of," and so on. Having such a structure allows us to draw certain inferences on the fly. For example, being told that something is a gas turbine immediately tells us that this object is a rotating equipment that has blades, requires fuel, generates electricity, and so on. This can be used practically in many asset management tasks. Combined with the Bayesian network approach, it reveals causal links between parts of an asset.

3.4 Training and assessing a model

In training any model, we use empirical data to tune the model parameters such that the model fits best to the data. That data is called the **training data**. Having made the model, we want to know how good the model is. While it is interesting how well it performs on the training data, this is not the final answer. We are usually much more concerned by how the model will perform on data it has not seen during its own construction, the **testing data**.

Some training algorithms not only have the training data to tune the model parameters but, additionally, use a dataset to determine when the training should stop because no significant model performance improvement can be expected. It is common to use the testing data for this purpose. It then may become necessary to generate a third dataset, the **validation data**, for testing the model on data that was not used at all during training.

It is common practice therefore to split the original dataset into two parts. We generally use 70%–85% of the data for training and the remainder for testing. Fig. 3.15 illustrates this process.

The choice of which data goes into the training or testing datasets needs to be made carefully because both datasets should be representative and significant for the problem at hand. Often, these are chosen randomly, and so chances are that some unintended bias enters the choice. This gave birth to the idea of **cross-validation**, which is the current accepted standard for demonstrating a model's performance.

The original dataset is divided into several, roughly equally sized, parts. It is typical to use 5 or 10 such parts and they are called **folds**. Let's say that we are using N folds. We now construct N different training datasets and N different testing datasets. The testing datasets consist of one of the folds and the training datasets consist of the rest of the data. Fig. 3.16 illustrates this division.

We now go through N distinct trainings, resulting in N distinct models. Each model is tested using its own testing dataset. The

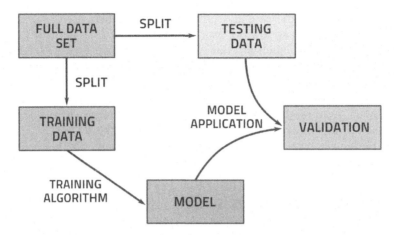

Fig. 3.15 The full dataset must be split into data to be used for model training and data to be used for testing the model's performance.

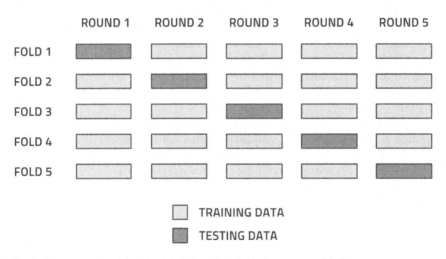

Fig. 3.16 An illustration of how to divide the full dataset into folds for cross-validation.

performance is now estimated to be the average performance over all the trained models. Generally, only the best model is kept but its performance has been estimated in this averaged fashion. This way of assessing the performance is more independent of the dataset bias than assessing it using any one choice of testing dataset and so it is the expected performance estimation method.

Having done all this, we could get the idea that we could use all N models, evaluate them all in each real case, and take the average of them all to be the final output of an overarching model. This is a legitimate idea that is called an **ensemble model**. The ensemble model consists of several fully independent models that may or may not have different architectures and that are averaged to yield the output of the full model. These methods are known to perform better on many tasks as compared to individual models, at least in the sense of variance. However, it comes at the expense of resources as several models must not only be trained but maintained and executed in each case. Especially for industrial applications, that often require real-time model execution, this may be prohibitive.

It is important to mention that the model and the training algorithm usually have some fixed parameters that a data scientist must choose prior to training. An example is the number of layers of a neural network and the number of neurons in each layer. Such parameters are called **hyperparameters**. It is rare indeed that we know what values will work best in advance. In order to achieve the best overall result, we must perform **hyperparameter tuning**. Often this is done in a haphazard trial-and-error method by hand because we do not want to try too many combinations. After all, we should ideally perform the full cross-validation for each choice and each one of those involves several model trainings. All this training takes time. It is possible however to perform automated hyperparameter tuning by using an optimization algorithm. If model performance is very important and we have both time and resources to spare, hyperparameter tuning is a very good way to improve model performance. Once this has been exhausted, there is little more one can do.

3.5 How good is my model?

On a weekly basis, we read of some new record of accuracy of some algorithm in a machine learning task. Sometimes it's image classification, then it's regression, or a recommendation engine. The difference in accuracy of the best model or algorithm to date from its predecessor is shrinking with every new advance.

A decade ago, accuracies of 80% and more were already considered good on many problems. Nowadays, we are used to seeing 95% already in the early days of a project. Getting (1) more sophisticated algorithms, (2) performing laborious hyperparameter tuning, (3) making models far more complex, (4) having access to vastly larger datasets, and last—but certainly not least—(5) making use of incomparably larger computer resources has driven accuracies to 99.9% and thereabouts for many problems.

Instinctively, we feel that greater accuracy is better and all else should be subjected to this overriding goal. This is not so. While there are a few tasks for which a change in the second decimal place in accuracy might actually matter, for most tasks this improvement will be irrelevant—especially given that this improvement usually comes at a heavy cost in at least one of the earlier five dimensions of effort.

Furthermore, many of the very sophisticated models that achieve extremely high accuracies are quite brittle and thus vulnerable to unexpected data inputs. These data inputs may be rare or forbidden in the clean datasets used to produce the model, but strange inputs do occur in real life all the time. The ability for a model to produce a reasonable output even for unusual inputs is its **robustness** or **graceful degradation**. Simpler models are usually more robust and models that want to survive the real world must be robust.

A real-life task of machine learning sits in between two great sources of uncertainty: life and the user. The data from life is inaccurate and incomplete. This source of uncertainty is usually so great that a model accuracy difference of tens of a percent may not even be measurable in a meaningful way. The user who sees the result of the model makes some decision on its basis and we know from a large body of human-computer interaction research that the user cares much more about how the result is presented, than the result itself. The usability of the interface, the beauty of the graphics, the ease of understanding and interpretability count more than the sheer numerical value on the screen.

In most use cases, the human user will not be able to distinguish a model accuracy of 95% from 99%. Both models will be considered "good" meaning that they "solve" the underlying problem that the model is supposed to solve. The extra 4% in accuracy are never seen but might have to be bought by many more resources both initially in the model-building phase as well as in the ongoing model execution phase. This is the reason we see so many prize-winning algorithms from competitions never being used in a practical application. They have high accuracy but either this high accuracy does not matter in practice or it is too expensive (complexity, project duration, financial cost, execution time, computing resources, etc.) in real operations.

We must not compare models based on the simplistic criterion of accuracy alone but measure them in several dimensions. We will then achieve a balanced understanding of what is "good enough" for the practical purpose of the underlying task. The outcome in many practical projects is that we are done much faster and with less resources. Machine learning should not be perfectionism but pragmatism.

3.6 Role of domain knowledge

Data science aims to take data from some domain and come to a high-level description or model of this data for practical use in solving some challenge in that domain. How much knowledge about the domain does the data scientist have to do a good job?

Before starting on a data science project, someone must define (a) the precise domain to focus on, (b) the particular challenge to be solved, (c) the data to be used, and (d) the manner in which the answer must be delivered to the beneficiary. All four of these aspects are not data science in themselves but have significant impact on both the data science and the usefulness of the entire effort. Let's call these aspects the **framework** of the project.

While doing the data science, the data must be assessed for its quality: precision, accuracy, representativeness, and significance.

- Precision: How much uncertainty is in a value?
- Accuracy: How much deviation from reality is there?
- Representativeness: Does the dataset reflect all relevant aspects of the domain?
- Significance: Does the dataset reflect every important behavior/dynamic in the domain?

In seeking a high-level description of the data, be it as a formulaic model or some other form, it is practically expedient to be guided by existing descriptions that may only exist in textual, experiential, or social forms, i.e., in forms inaccessible to structured analysis. In real projects we find that data science often finds (only) conclusions that are trivial to domain experts or does not find a significant conclusion at all. Incorporating existing descriptions will prevent the first and make the second apparent a lot earlier in the process.

It thus becomes obvious that domain knowledge is important both in the framework as well as the body of a data science project. It will make the project faster, cheaper, and more likely to yield a useful answer.

This situation is beautifully illustrated by the famous elephant parable. Several blind persons, who have never encountered an elephant before, are asked to touch one and describe it. The descriptions are all good descriptions, given the experience of each person, but they are all far from the actual truth, because each person was missing other important data, see Fig. 3.17.

This problem could have been avoided with more data or with some contextual information derived from existing elephantine descriptions. Moreover, the effort might be better guided if it is clear what the description will be used for.

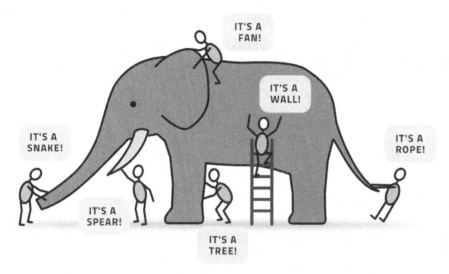

Fig. 3.17 The parable of the six blind people trying to describe an elephant by touching part of it.

A domain expert typically became an expert by both education and experience in that domain. Both imply a significant amount of time spent in the domain. As most domains in the commercial world are not freely accessible to the public, this usually entails a professional career in the domain. This is a person who could define the framework for a data science project, as they would know what the current challenges are and how they must be answered to be practically useful, given the state of the domain as it is today. The expert can judge what data is available and how good it is. The expert can use and apply the deliverables of a data science project in the real world. Most importantly, this person can communicate with the intended users of the project's outcome. This is crucial, as many projects end up being shelved because the conclusions are either not actionable or not acted upon.

A data scientist is an expert in the analysis of data. Becoming such an expert also requires a significant amount of time spent being educated and gaining experience. Additionally, the field of data science is developing rapidly, so a data scientist must spend considerable time keeping up with innovations. This person decides which of the many available analysis methods should be used in this project and how these methods are to be parametrized. The tools of the trade (usually software) are familiar to this person, and he or she can use them effectively. Model quality and goodness of fit are evaluated by the data scientist. Communication with

data scientist

domain expert

1. Education in data
2. Experience in data
3. Availability of methods
4. Configuration of tools
5. Model quality
6. Communication with technical staff

1. Education in domain
2. Experience in domain
3. Application of tools
4. Data availability
5. Data quality
6. Communication with intended users

Fig. 3.18 The difference in roles between a data scientist and a domain expert.

technical persons, such as mathematicians, computer scientists, and software developers, can be handled by the data scientist (Fig. 3.18).

The expectation that a single individual would be capable of both roles is unrealistic, in most practical cases. Just the requirement of time spent, both in the past as well as regular upkeep of competence, prohibits dual expertise. In some areas it might be possible for a data scientist to learn enough about the domain to make a good model, but assistance would still be needed in defining the challenge and communicating with users, both of which are highly nontrivial. It may also be possible for a domain expert to learn enough data science to make a reasonable model, but probably only when standardized tools are good enough for the job (Fig. 3.19).

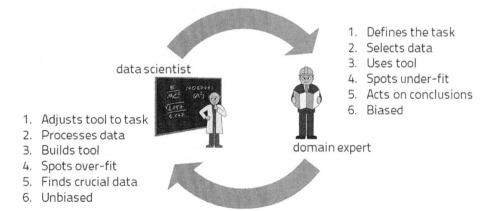

data scientist

domain expert

1. Adjusts tool to task
2. Processes data
3. Builds tool
4. Spots over-fit
5. Finds crucial data
6. Unbiased

1. Defines the task
2. Selects data
3. Uses tool
4. Spots under-fit
5. Acts on conclusions
6. Biased

Fig. 3.19 A project is much more likely to succeed if a domain expert and data scientist collaborate.

If there are two individuals, they can get excellent results quickly by good communication. While the domain expert (DE) defines the problem, the data scientist (DS) chooses and configures the right toolset to solve it. The representative, significant, and available data are chosen by the DE and processed by the DS. The DS builds the tool (that might involve programming) for the task given the data, and the DE uses the tool to address the challenge. Having obtained a model, the DS can spot overfitting, where the model has too many parameters so that it effectively memorizes the data, leading to excellent reproduction of training data, but poor ability to generalize. The DE can spot underfitting, where the model provides too little accuracy or precision to be useful or applied in the real world. The DS can isolate the crucial data in the dataset needed to make a good model; frequently this is a small subset of all the available data. The DE then acts on the conclusions by communicating with the users of the project and makes appropriate changes. The DS approaches the project in an unbiased way, looking at data just as data. The DE approaches the project with substantial bias, as the data has significant meaning to the DE, who has preformed hypotheses about what the model should look like. It is important to note that bias, in this context, is not necessarily a detriment to the effort.

It is instructive to think what the outcome can be if we combine a certain amount of domain knowledge with a certain level of data science capability. The following vision is the author's personal opinion but is probably a reasonable reflection of what is possible today.

First, for the sake of this discussion, let's divide domain knowledge into four levels: (1) Awareness is the basic level where we are aware of the nature of the domain. (2) Foundation is knowing what the elements in the domain do, equivalent to a theoretical education. (3) Skill is having practical experience in the domain. (4) Advanced is the level where there is little left to learn and where skill and knowledge can be provided to other people, i.e., this person is a domain expert.

Similarly, we can divide up data science into five levels: (1) Data is where we have a table or database of numbers. We can do little more than draw diagrams with this. (2) Information is where we have descriptive statistics about the available data, such as correlations and clusters. (3) Knowledge is when we have some static models. (4) Understanding is when we have dynamic models. The distinction between static and dynamic models is whether the model incorporates the all-important variable of time. A static model makes a statement about how one part of the process affects another, whereas a dynamic model makes

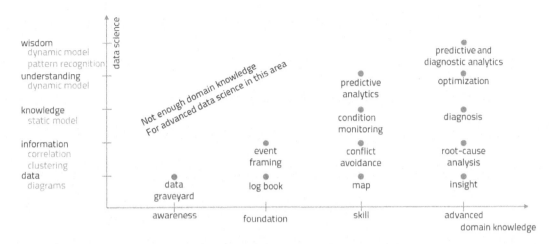

Fig. 3.20 More is possible if sophisticated domain knowledge is combined with deep data science expertise.

additional statements about how the past affects the future. (5) Wisdom is when we have both dynamic models and pattern recognition, for now we can tell what will happen when.

In Fig. 3.20 are several technologies that exist today ordered by the level of data science that they represent and the amount of domain knowledge that was necessary to create them. There are no technologies in the upper left of the diagram because one cannot make such advanced data science with so little domain knowledge.

In conclusion, data science needs domain knowledge. As it is unreasonable to expect any one person to fulfill both roles, we are necessarily looking at a team effort.

3.7 Optimization using a model

We train a machine learning model by adjusting its parameters such that its performance—the least-squares difference between model output and expected output—is a minimum. Such a task is called **optimization** and methods that do this are called optimization algorithms. Every training algorithm for a model is an example of an optimization algorithm. Usually these algorithms are highly customized to the kind of machine learning model that they deal with. We will not treat these methods as that level of detail is beyond the scope of the book.

There are also some general-purpose optimization algorithms that work on any model. These methods can be used on the machine learning model that was just made. If we have a model for the efficiency of a gas turbine, for example, as a function of a multitude of input variables, then we can ask: How should the input variables be modified in order to achieve the maximum possible efficiency? This is a question of **process optimization** or **advanced process control**, which is also sometimes—strangely—called **statistical process control** (Coughanowr, 1991; Qin and Badgwell, 2003).

For the sake of illustration, let's say that we have only two manipulated variables that we can change in order to affect the efficiency. Fig. 3.21 is a contour plot that illustrates the situation. The horizontal and vertical directions are the two manipulated variables. The efficiency resulting from these is displayed in contour lines just like on a topographical map. The points labeled A, B, and C are the peaks in the efficiency.

Any practical problem has boundary conditions, i.e., conditions that prevent us from a particular combination of manipulated variables. These usually arise from safety concerns, physical constraints, or engineering limitations and are indicated by the dashed lines. The peak labeled A, for instance, is not allowed due to the boundary conditions.

The starting point in the map is the current combination of manipulated variables. Whatever we do is a change away from this point. What is the best point? It is clearly either B or C. The resolution of the map does not tell us but in practice these two efficiencies

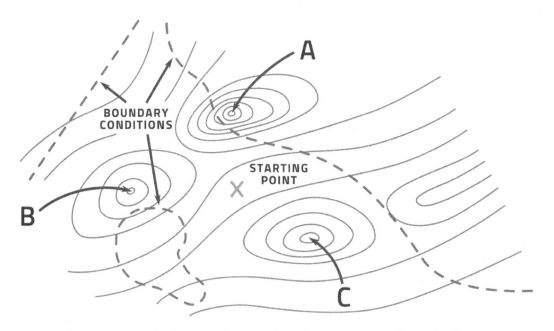

Fig. 3.21 A contour plot for a controlled variable like efficiency in terms of two manipulated variables.

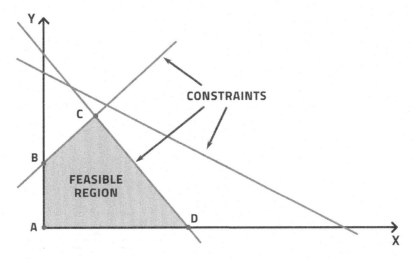

Fig. 3.22 Linear optimization illustrated as a region bounded by constraints with the optimal points necessarily on the border.

may be quite close to one another. This is like a car's GSP choosing one route over another because it computes one path to be a few seconds shorter. Whichever point we choose to head for, the next task is to find the right path. In terms of efficiency, this is perpendicular to the contour lines, i.e., the steepest ascent.

Computing these actions in advance, with the appropriate amount of time that is needed in between variable changes, is the purpose of the optimization. Then the actions can be realized in the physical process and the efficiency should rise as a result.

There is a special case of optimization where the objective is a linear function of all the measurable variables. This is known as **linear programming** and is displayed in Fig. 3.22. The boundary conditions, or **constraints**, define a region that contains all the points allowed. We can prove that the best solution is on the boundary of the region and solving such problems is relatively simple and fast. Linear optimization problems are routinely solved in the oil and gas industry, particularly in financial optimization, e.g., how much of each refinery end product to make during any 1 week in a refinery.

So, we have a model $y = f(\underline{x})$ where the function $f(\cdots)$ is potentially a highly complex machine learning model, $\underline{x}$ is potentially a long vector of various measurements, and y is the target of the optimization, i.e., the quantity that we want to maximize. Of the elements of $\underline{x}$, there will be some that the operators of the facility can change at will. These are called **set points** because we can set them. There will also be values that are fixed by powers that are completely

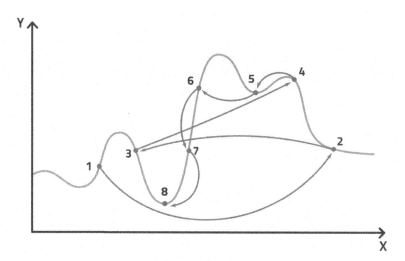

Fig. 3.23 Simulated annealing illustrated as a sequence of jumps, sometimes for the better and sometimes for the worse, eventually converging on the global minimum.

beyond our control, e.g., the weather, market prices, raw material qualities, and other factors determined outside our facility. These are fixed by the outside world and so correspond to **constraints**.

Now we can ask: What are the values of the set points, given all the boundary conditions and constrained values, that lead to the largest possible value of y? This is the central question of optimization. Let us call any setting of the set points that satisfies all boundary conditions, a **solution** to the problem. We then seek the solution with the maximum outcome in y.

One of the most general ways of answering this question is the method of **simulated annealing**. This method starts from the current setting of the set points. We then consider a random change in the vector of set points such that this second point is still a solution. If the new point is better than the old, we accept the change. If it is worse, we accept it with a probability that starts off being high and that gradually decreases over the course of the iterations. We keep on making random changes and going through this probabilistic acceptance scheme. Once the value of the optimization target does not change anymore over some number of iterations, we stop the process and choose the best point encountered over the entire journey. See Fig. 3.23 for an illustration of a possible sequence of points in search for the global minimum.

This method provably converges to the best possible point (the global optimum). Even if we cut off the method after a certain finite amount of resources (such as computation time or number of transitions) has been spent on it, we can rely on the method having produced a reasonable improvement relative to the resources spent.

There are a few fine points of simulated annealing and there is much discussion in the literature on how to calculate the probability of accepting a temporary change for the worse, how to judge if the method has converged, how to generate the new trial point, and so on (Bangert, 2012).

3.8 Practical advice

If you want to perform a machine learning project in real life, here is some advice on how to proceed.

First, all the steps outlined in Section 2.11 should be followed so that you have a clean, representative, significant dataset and you are clear on what you want as the final outcome and how to measure whether it is good enough for the practical purpose.

Second, as mentioned in Section 3.6 it is a good idea to conduct such a project with more than one person. Without going into too much detail here, you may need a small team consisting of one or several of these persons: an equipment operator, a maintenance engineer, a process engineer, a process control specialist, a person from the information technology (IT) department, a sponsor from higher management, a data scientist, and a machine learner. From practical experience, it is virtually impossible to do such a project by oneself.

Third, select the right kind of mathematical modeling approach to solve your problem. Several popular ones were outlined earlier. There are more but these represent most models applied in practice. You may ask yourself the following questions:

1. Do you need to calculate values at a future time? You will probably get the best results from a recurrent neural network such as an LSTM as it is designed to model time dependencies.
2. Do you need to calculate a continuous number in terms of others? This is the standard regression problem and can be reliably solved using deep neural networks.
3. Do you need to identify the category that something belongs to? This is classification and the best general classification method is random forest.
4. Do you need to compute the likelihood that something will happen (risk analysis) or has happened (root-cause analysis)? Bayesian networks will do this very nicely.
5. Do you need to do several of these things? Then you will need to make more than one model.

Fourth, every model type and training method has settings. These are the hyperparameters as the model coefficients are usually called the parameters. Having selected the model type, you will now have to set the hyperparameters. This can be done by throwing resources at the issue and doing an automated hyperparameter tuning in which a computer tries out many of them and selects the best. If you have the time to do this, do it. If not, you will need to think the choices through carefully. The main consideration is the bias-variance-complexity trade-off that mainly concerns those hyperparameters that influence how many parameters the model has.

Fifth, perform cross-validation as described in Section 3.4 to calculate all the performance metrics you have chosen correctly. This will get you an objective assessment of how good your model is as described in Section 3.5.

Sixth, if your goal was the model itself, then you now have it. However, if your goal is to derive some action from the model, you now need to compute that action using the model and this usually requires an optimization algorithm as described in Section 3.7. Several such are available but simulated annealing has several good properties that make it an ideal candidate for practical purposes.

Seventh, communicate your results to the rest of the team and to the wider audience in a way that they can understand it. You may need the domain experts to help you do this properly. This audience may need to change their long-established ways based on your results, and they may not wish to do so. The implied change management in this communication is the most difficult challenge of the entire project. In my personal experience this is the cause of failure in almost all failed data science projects in the industry, which is why we will address this at length in other chapters of this book.

References

Agrawal, A., Gans, J., Goldfarb, A., 2018. Prediction Machines. Harvard Business Review Press, Brighton.

Bangert, P.D., 2012. Optimization for Industrial Problems. Springer, Heidelberg.

Bishop, C.M., 2006. Pattern Recognition and Machine Learning. Springer, Heidelberg.

Breiman, L., Friedman, J.H., Olshen, R.A., Stone, C.J., 1984. Classification and Regression Trees. Chapman & Hall, London.

Cortes, C., Vapnik, V.N., 1995. Support-vector networks. Mach. Learn. 20, 273–297.

Coughanowr, D.R., 1991. Process Systems Analysis and Control, third ed. McGraw-Hill, New York City.

Domingos, P., 2015. The Master Algorithm. Basic Books, New York City.

Ebrahimipour, V., Yacout, S., 2015. Ontology Modeling in Physical Asset Integrity Management. Springer, Heidelberg.

Gelman, A., Carlin, J.B., Stern, H.S., Rubin, D.B., 2004. Bayesian Data Analysis. Chapman & Hall, London.

Gers, F., 2001. Long Short-Term Memory in Recurrent Neural Networks. (Ph.D. thesis)Ecole Polytechnique Federale de Lausanne.

Goodfellow, I., Bengio, Y., Courville, A., 2016. Deep Learning. MIT Press, Cambridge.

Hagan, M.T., Demuth, H.B., Beale, M., 1996. Neural Network Design. PWS, Boston.

Hochreiter, S., Schmidhuber, J., 1997. Long short-term memory. Neural Comput. 9, 1735–1780.

Kohonen, T., 2001. Self-Organizing Maps, third ed. Springer, Heidelberg.

MacKay, D.J.C., 2003. Information Theory, Inference, and Learning Algorithms. Cambridge University Press, Cambridge.

Mitchell, T.M., 2018. Machine Learning. McGraw-Hill, New York City.

Qin, S.J., Badgwell, T.A., 2003. A survey of industrial model predictive control technology. Control. Eng. Pract. 11, 733–764.

Scutari, M., Denis, J.B., 2015. Bayesian Networks. Chapman & Hall, London.

Seiffert, U., Jain, L.C., 2002. Self-Organizing Neural Networks. Physica Verlag, Heidelberg.

Yu, Y., Si, X., Hu, C., Zhang, J., 2019. A review of recurrent neural networks: LSTM cells and network architectures. Neural Comput. 31, 1235–1270.

4

Introduction to machine learning in the power generation industry

Patrick Bangert[a,b]
[a]*Artificial Intelligence, Samsung SDSA, San Jose, CA, United States*
[b]*Algorithmica Technologies GmbH, Bad Nauheim, Germany*

Chapter outline

This chapter will attempt to provide an overview over some of the practical applications that machine learning has found in power generation. The aim of the chapter is twofold: First, it is to show that there are many applications that are realistic and have been carried out on real-world assets, i.e., machine learning is not a dream. Second, the status of machine learning in power is in its early days as the applications are specialized and localized. It must be stated clearly that the majority of the studies performed have been done at universities and that the applications fully deployed in commercial power companies are the exception. This chapter makes no attempt at being complete or even representative of the work done. It just provides many starting points for research on use cases and presents an overview. There are some use cases that attract a vast number of papers and this chapter will present such use cases with just one or a few exemplary papers chosen at random.

This chapter can be used as a starting point to searching for literature on a specific application case and as an overview over the fact that there are a great many different use cases in power

Machine Learning and Data Science in the Power Generation Industry. https://doi.org/10.1016/B978-0-12-819742-4.00004-4
© 2021 Elsevier Inc. All rights reserved.

generation for machine learning both in terms of the type of use as well as the area of application. Some review papers have already appeared on this topic that also provide a good overview (Hatziargyriou, 2001; Cheng and Tao, 2019; Ge et al., 2017; Ma et al., 2019).

4.1 Forecasting

Many applications focus on the forecasting of a time series into the near future. Often, the forecast is just for hours or a few days. Considering the status of renewable energy and its reliance on the sun shining and the wind blowing, **forecasting solar radiance and wind** are critical to knowing how much renewable power we can expect. Both of these cases have attracted a great many papers and it is discussed in detail in this book in Chapter 8. For some more information on forecasting wind, see Foley et al. (2012) and Zhang et al. (2016) and for forecasting solar radiation, see Sharma et al. (2011) and Akhter et al. (2019). Rather than forecasting the power output, one may focus on forecasting the power ramps, i.e., events during which the power output changes dramatically (Gan and Ke, 2014). One may also forecast ocean wave conditions for tidal power plants (James et al., 2018).

Even the power output of a base load power plant is not fully stable and needs to be forecasted in the short term to make sure that promises made to the grid can be fulfilled (Tüfekci, 2014). The security of a grid can be forecasted based on the available resources to make a risk assessment (Karim et al., 2018).

In any electricity grid, production must match consumption at any time. Adding to the difficulty of forecasting the availability of solar and wind power, the **consumer demand** must be forecasted as well. There are many papers on this topic as well, although not nearly as many as forecasting solar and wind power generation. These include Paterakis et al. (2017), Bonetto and Rossi (2017), and Wood et al. (2014). Not only do we need to forecast the demand for electricity but also the demand for district heating as many power plants provide both electricity and heat (Johansson et al., 2017; Provatas, 2014).

From the point of view of a single power plant, it is important to know whether the **load dispatcher** will require the plant to increase or decrease its load in the near future. This has been looked at as a forecasting problem in order to assist operators to plan properly and deduce appropriate response strategies (Staudt et al., 2018). One may also look at this as a **profit maximization** task and attempt to schedule power production plans one day ahead of time (Latha, 2018; Keerthisinghe et al., 2018).

One may even want to predict the voltage regulator tap changes in the context of distributed generation (Blakely et al., 2018).

4.2 Predictive maintenance

The general topic of **predictive maintenance** features largely in the popular literature and discussion in the industry. In that phrase, the word "predictive" means different things to different people.

On the one hand, it means forecasting an event. That is, we calculate at what time in the future something will happen and alert the operator to that future fact. In that sense, prediction is a synonym for forecast.

On the other hand, it sometimes means recognizing an event. That is, we calculate that there is something worrisome going on right now. In that sense, prediction is a synonym for computation. This use case is often called **anomaly detection**. For marketing reasons, many products offered under the label predictive maintenance actually deliver anomaly detection or some form of advanced **condition monitoring**. Anomaly detection can be done in an unsupervised way as discussed in a whole book (Aldrich and Auret, 2013).

It is worthwhile to be aware of these different usages of the term so that one can be careful when reading texts about some technologies. In the context of this book, predictive maintenance will be used in the sense of forecasting something in the future whereas anomaly detection will be used in the context of detecting something going on at the present moment. The two are clearly very distinct use cases because predictive maintenance allows you to prepare, plan, and execute an action before the forecasted moment comes to pass, i.e., it allows you to prevent the failure event. For this reason, the forecast must be made a macroscopic amount of time into the future—a forecast by several seconds is usually useless.

Predictive maintenance studies have been done about many different types of equipment relating to the power industry. Following, we present a list of samples for forecasting without any attempt at being comprehensive (Bangert, 2017a, 2019):

- Review article on predictive maintenance in power generation (Cline et al., 2017).
- Ampacity in conductor lines (Aznarte and Siebert, 2017).
- Power transformers (Sica et al., 2015).
- Blade tear on a steam turbine (Bangert, 2010).
  Here are some papers for anomaly detection (Kroll et al., 2014) (Bangert, 2017c, 2018):
- Solar panel defect detection (Lee and Lee, 2016).
- Short circuit detection in DC microgrids (Almutairy and Alluhaidan, 2017).
- Diagnosis of series arc faults in DC systems (Telford et al., 2017).
- Cavitation in a centrifugal pump (Dutta et al., 2018).

- General fault detection in power systems (Tokel et al., 2018).
- Fuel consumption by diesel generators (Mulongo et al., 2020).
- Fault detection in circulating water systems (Aziz et al., 2013).
- Gear fault detection (Yang et al., 2011).
- Wind turbines (Stetco et al., 2019; Leahy et al., 2016; Cui et al., 2018; Bangert, 2015b).
- Microgrid faults (Mishra and Rout, 2018).
- Frequency estimation (Karapidakis, 2007).
- General fault detection in mechatronic systems (Papakonstantinou et al., 2014).
- Gas circulator units (Costello et al., 2017).
- Grid structure (Uçar et al., 2016).
- General reliability management (Duchesne et al., 2017).
- Centrifugal water pumps (Chakravarthy et al., 2019).
- Power system relays (Fang and Zhang, 2019).
- Photovoltaic systems (De Benedetti et al., 2018).
- Radial fan impeller (Zenisek et al., 2019).
- Nuclear infrastructure (Gohel et al., 2020).
- Using infrared thermography (Ullah et al., 2017).

With some insight as to which equipment is in what state and can be expected to run, a plant may want to **schedule planned outages** in the best possible way so that it can conduct all the necessary maintenance measures in that same outage. It is undesirable to maintain something too early. On the other hand, one also does not want something to fail a few weeks after a major outage. Good scheduling as well as good planning of measures is critical in keeping the plant running in between major planned outages. These events can be scheduled and planned by models (Dalal et al., 2019). Dynamic decision making in the context of predictive maintenance can also be supported by models (Susto et al., 2015).

In studying predictive maintenance in all its forms, it is important to distinguish a **drifting sensor** from a real change in signal. All physical sensors will drift over time and need to occasionally be recalibrated. This effect may lead either to seemingly abnormal conditions, or happen so gradually as to redefine what a model might consider normal. In any case, telling the difference is important and this too can be done with analytics (Zenisek et al., 2019).

After anomaly detection is done, we naturally ask what type of anomaly it is. This is asking for the failure mode, the physical location, the type of problem, or generally for some sort of description or **diagnosis** of what is going on. Ultimately, this would lead to a maintenance measure plan in which certain individuals will be sent on site with tools and spare parts to perform a task. We must know the task in order to plan for the people, tools, and parts to be on hand at the right time. The issue of diagnosis is important (Jahnke, 2015).

4.3 Integration into the grid

As the power generation industry is necessarily a network of many parts, individual models must be integrated into that larger whole. This effort is often called the **smart grid** and is studied in its own right as for example in Perera et al. (2014).

As many providers feed power into the grid at random times (e.g., when the sun shines on your roof), a new problem called **islanding** is created. This occurs when distributed generators—such as wind turbines or solar panels—supply electricity to the grid even when the main utility is turned off. A worker will assume that the grid is power free and might perform maintenance work. Detecting such islands is thus a major safety task and a difficult problem in its own right (Alam et al., 2017; Alshareef et al., 2014). Islanding is a particular issue in the context of combined heat and power plants (Shao et al., 2018).

The advent of solar, wind, and biogas power has made it possible for private households to become power generators and contribute directly to the grid. This presents not only a physical challenge to the grid but also an economic one as the number of legal entities that supply power has increased manyfold. In order to lessen at least the economic challenges and perhaps also some physical ones, one can aggregate several household generators into a larger collective known as a **virtual power plant**. These are interesting targets for modeling their flexibility, i.e., their matching between supply and demand (MacDougall et al., 2016).

As a member of a grid, any power generator has to plan power production based on the physical and economic needs of the grid and negotiations with the load dispatcher. Scheduling production is thus a difficult task that can be profitably treated by modeling (Watanabe et al., 2018). In the same context, it is necessary to negotiate with the grid as often as every 15 minutes regarding load and price. This can be done by software-based broker agents driven by reinforcement learning approaches (Peters et al., 2013).

4.4 Modeling physical relationships

Machine learning can assist in modeling physical relationships between equipment parts. This can allow us to better understand these equipments without having to perform a physics-based or first principles model, which is usually much more work to produce.

For example, the relationship between wind speed and the power produced by **wind turbines** is nonlinear and therefore requires modeling to compute properly (Marvuglia and Messineo, 2012). When looking at wind turbines, it is instructive to remember that they are mostly deployed as a collection of wind turbines closely spaced in an area. As the wind interacts with the first wind turbines, the other wind turbines receive less wind and so extract less power from the wind. Modeling this **wake** of a wind turbine has a significant influence over the management of a wind farm (Ti et al., 2020).

The proper workings of **coal mills** can be modeled and therefore their operation and design improved (Zhang et al., 2002).

Gas turbines and compressors require sophisticated models to control, operate, maintain, and design (Liu and Karimi, 2020). In particular, we may be interested in the vibration behavior of gas turbines as the vibration allows many interesting conclusions to be drawn about the health status and remaining useful life of the turbine (Zárate et al., 2020).

Arguably the most important technology for the medium-term future of the power generation industry is the **battery**. With more and more intermittent generation from renewable sources, it will become ever more important to be able to store electricity. Some will need to be stored for only short periods of time to buffer the grid but it will become more necessary to store power for several hours to 1–2 days in order to overcome the intermittent nature of solar power (Hariharan et al., 2018) and wind power (Zhang and Li, 2013). Investigating new technologies for batteries as well as the control of large arrays of battery cells into large batteries is important. Example papers include (Hu et al., 2016; Wang, 2019). Once there is a significant battery infrastructure, this will influence grid management significantly, which will have to be managed appropriately (Ishizaki et al., 2017).

As we are all painfully aware however, batteries deplete and age over time. This behavior must be modeled and controlled as much as possible (Severson et al., 2019; Henri et al., 2018). More crucially perhaps, it seems clear that the currently dominating lithium-ion battery technology cannot provide the capacity at scale that would be needed for these deployments. On the one hand, the raw materials of lithium, cobalt, and so on are too scarce and on the other hand these elements leave toxic waste (Prior et al., 2013). An alternative must be found that is benign, cheap, robust, long-lasting, and the materials for which are available in plenty. The search is on but machine learning is not contributing much to this effort, to date at least.

While generating electricity, many plants directly produce **pollution and greenhouse gases** such as NOx. Modeling how this is produced and then using the model to control and mitigate that production is an important application to reduce the impact of power generation on the environment (Tan et al., 2016).

4.5   Optimization and advanced process control

As machine learning can learn and represent the reaction of a mechanical system to external inputs, the model can be turned around. That is, we could stipulate what we desire the outcome to be and ask what the inputs need to be to bring that about. In practice: How should we change the set points of the process in order to maximize some measure of performance or economics?

These methods are sometimes called **process optimization** and are often included in the term **advanced process control**. Strictly speaking, process control has as its goal the keeping constant of some quantity but we can be more general and also include the maximization or minimization of a quantity. In doing this, we require two mathematical technologies: a model and an optimization method to invert the model(Bangert, 2014, 2015a, 2017b). For example, the advanced process control of a CHP plant may be optimized to deliver over 1% absolute extra efficiency to the plant just by changing some set-points (Bangert and Czernitzky, 2010).

Applications include the optimization of district heating (Idowu et al., 2014). Filtration systems occasionally foul and need to be cleaned. This process can be controlled to some extent, which has been studied with machine learning (Bagheri et al., 2019). Wind turbines need to be controlled in reaction to changing wind conditions (Chehaidia et al., 2020). Biogas power plants can also be controlled with models (Oluwaseun et al., 2019). So many papers have appeared just on the issue of controlling rotor angle stability that a review paper has appeared (Yousefian and Kamalasadan, 2017).

Buildings are highly complex electricity users and proper control systems can significantly reduce electricity usage as well as electricity cost if the usage is properly distributed. Building control systems have been designed based on learning principles (Zhou and Zheng, 2020).

Grids need to be controlled to ensure stability and optimal flow (Karagiannopoulos et al., 2019). In particular, the smart inverter in these grids is a weak point that can benefit from model-based control (Garg, 2018). In fact, an electricity grid can only become what is known as a smart grid by including several elements of artificial intelligence and analytics (Azad et al., 2019). Grids rely on optimal power flow that can be deduced and assured through modeling (Guha et al., 2019; Dobbe et al., 2020).

Many plants are still controlled by proportional-integral-differential (PID) controllers. These can be augmented by neural networks to improve performance without shifting the basic paradigm completely to machine learning. This has been investigated in the context of hydroelectric plants (Zhang and Wang, 2002).

4.6 Consumer aspects

Electricity demand is volatile and increases over time as people become more numerous and consume more resources. One way to alleviate the pressure on the grid is to use **home appliances** efficiently. It is possible to identify a home appliance by its unique load signature and thus use appliances at the right time (Khan et al., 2018). This also allows users to identify the right appliances for replacement with more efficient versions. This can be used in various ways around the home in managing the solar panels on the roof, the electric car in the garage, the heating system of the house, and so on (Shahriar and Rahman, 2015).

In encouraging users to change their **behavior**, we must take account of the willingness and ability of the user to comply and alter their usage patterns. Human behavior, at least statistically, is also amenable to numeric description and machine learning analysis (Bei et al., 2011). The consumer can be guided in decision making of the smart home by various decision support systems (Li and Jayaweera, 2015). From the view of power generators, we may wish to model the consumer's behavior in order to better provide for them (Gariba and Pipaliya, 2016).

Buildings are a significant consumer of electricity and so it is instructive to model the energy demand of buildings. This book addresses this issue in detail in Chapter 9 in the case of a hospital. Numerous papers are published on modeling, forecasting, and reducing the energy demanded by buildings and their major electricity consuming systems (Jurado et al., 2015; Pallonetto et al., 2019). Forecasting the heat consumption in buildings is the other side of the coin and should be studied in tandem (Idowu et al., 2016).

4.7 Other applications

In engineering power systems and equipment, designers frequently rely on certain material properties. When these reach their limits, one seeks novel materials that extend these properties. Machine learning has been successfully used to envision and design novel materials to solve engineering problems (Umehara et al., 2019).

Most power generation facilities require large amounts of water. There is a steam cycle in which water is transformed into steam that drives steam turbines. There is cooling water that cools down parts of the heated process. Even before reaching the power plant, water is involved in mining the fuels (gas, coal, oil, uranium), refining the fuels (oil, gas, uranium), as well as their transportation to the plant (coal through slurries). Pollution control systems require lots of water to operate. Machine learning has been used to address the issues that arise from the copious water demand of power generation (Zaidi et al., 2018).

While machine learning in power generation is getting more and more attention, it should be noted that some machine learning systems are brittle in that they do not generalize well beyond the training dataset. In this context, such models are vulnerable to so-called **adversarial designs**. These artificially constructed input data points are purposefully made to confuse the model into providing a wrong output. Such designs can be used maliciously as a form of attack on a system controlled by machine learning (Chen et al., 2018).

More generally, power system security on a physical level must be maintained and assured at all times (Wehenkel, 1997). Beyond the physical, when plants and grids are becoming ever more digitized, we must pay attention to cybersecurity as well (Hossain et al., 2019).

References

Akhter, M.N., Mekhilef, S., Mokhlis, H., Mohamed, N., 2019. Review on forecasting of photovoltaic power generation based on machine learning and metaheuristic techniques. IET Renew. Power Gener. 13 (7), 1009–1023.

Alam, M.R., Muttaqi, K.M., Bouzerdoum, A., 2017. Evaluating the effectiveness of a machine learning approach based on response time and reliability for islanding detection of distributed generation. IET Renew. Power Gener. 11 (11), 1392–1400.

Aldrich, C., Auret, L., 2013. Unsupervised Process Monitoring and Fault Diagnosis With Machine Learning Methods. Springer, Heidelberg.

Almutairy, I., Alluhaidan, M., 2017. Fault diagnosis based approach to protecting DC microgrid using machine learning technique. Procedia Comput. Sci. 114, 449–456.

Alshareef, S., Talwar, S., Morsi, W.G., 2014. A new approach based on wavelet design and machine learning for islanding detection of distributed generation. IEEE Trans. Smart Grid 5 (4), 1575–1583. https://doi.org/10.1109/TSG.2013.2296598.

Azad, S., Sabrina, F., Wasimi, S., 2019. Transformation of smart grid using machine learning. In: 2019 29th Australasian Universities Power Engineering Conference (AUPEC), Nadi, Fiji, pp. 1–6. https://doi.org/10.1109/AUPEC48547.2019.211809.

Aziz, N.L.A.A., Yap, K.S., Bunyamin, M.A., 2013. A hybrid fuzzy logic and extreme learning machine for improving efficiency of circulating water systems in power generation plant. IOP Conf. Ser.: Earth Environ. Sci. 16.

Aznarte, J.L., Siebert, N., 2017. Dynamic line rating using numerical weather predictions and machine learning: a case study. IEEE Trans. Power Delivery 32 (1), 335–343. https://doi.org/10.1109/TPWRD.2016.2543818.

Bagheri, M., Akbari, A., Mirbagheri, S.A., 2019. Advanced control of membrane fouling in filtration systems using artificial intelligence and machine learning techniques: a critical review. Process. Saf. Environ. Prot. 123, 229–252.

Bangert, P., 2010. Two-Day Advance Prediction of a Blade Tear on a Steam Turbine of a Coal Power Plant. pp. 175–182 (in Link, M., 2010. Schwingungsanalyse & Identifikation. In: Proceedings of the Conference of 23./24.03.2010 in Leonberg, Germany. (VDI-Berichte No. 2093). ISBN:978-3180920931).

Bangert, P., 2014. Increasing the profitability of CHP plants by 1–2% using mathematical optimization. FuturENERGY 5, 21–24.

Bangert, P., 2015a. Increase overall fuel efficiency of power plants using mathematical modeling. In: Proceedings of the 58th Annual ISA POWID Symposium.

Bangert, P., 2015b. Failures of wind power plants can be predicted several days in advance. In: Proceedings of the 58th Annual ISA POWID Symposium.

Bangert, P., 2017a. Predicting and detecting equipment malfunctions. OTC Brasil. OTC-28109-MS.

Bangert, P., 2017b. Optimization of CHP and fossil fuels by predictive analytics. In: Proceedings of the 60th Annual ISA POWID Symposium.

Bangert, P., 2017c. Smart condition monitoring using machine learning. In: Proceedings of the 60th Annual ISA POWID Symposium.

Bangert, P., 2018. Algorithmica's machine learning enables predictive maintenance and optimization. Managing Aging Plants (April).

Bangert, P., 2019. Predicting and detecting equipment malfunctions using machine learning. SPE. 195149.

Bangert, P., Czernitzky, J., 2010. Increase of overall combined-heat-and-power (CHP) efficiency via mathematical modeling. In: Proceedings of the VGB Power-Tech e.V. Conference "Fachtagung Dampferzeuger, Industrie- und Heizkraftwerke 2010".

Bei, L., Gangadhar, S., Cheng, S., Verma, P.K., 2011. Predicting user comfort level using machine learning for smart grid environments. In: ISGT 2011, Anaheim, CA, pp. 1–6. https://doi.org/10.1109/ISGT.2011.5759178.

Blakely, L., Reno, M.J., Broderick, R.J., 2018. Decision tree ensemble machine learning for rapid QSTS simulations. In: 2018 IEEE Power & Energy Society Innovative Smart Grid Technologies Conference (ISGT), Washington, DC, pp. 1–5. https://doi.org/10.1109/ISGT.2018.8403323.

Bonetto, R., Rossi, M., 2017. Machine learning approaches to energy consumption forecasting in households. arXiv:1706.09648.

Chakravarthy, R., Bharadwaj, S.C., Padmanaban, S., Dutta, N., Holm-Nielsen, J.B., 2019. Electrical fault detection using machine learning algorithm for centrifugal water pumps. In: 2019 IEEE International Conference on Environment and Electrical Engineering and 2019 IEEE Industrial and Commercial Power Systems Europe (EEEIC/I&CPS Europe), Genova, Italy, pp. 1–6. https://doi.org/10.1109/EEEIC.2019.8783841.

Chehaidia, S.E., Abderezzak, A., Kherfane, H., Boukhezzar, B., Cherif, H., 2020. An improved machine learning techniques fusion algorithm for controls advanced research turbine (CART) power coefficient estimation. UPB Sci. Bull. Ser. C 82 (2), 279–292.

Chen, Y., Tan, Y., Deka, D., 2018. Is machine learning in power systems vulnerable? In: 2018 IEEE International Conference on Communications, Control, and Computing Technologies for Smart Grids (SmartGridComm), Aalborg, pp. 1–6. https://doi.org/10.1109/SmartGridComm.2018.8587547.

Cheng, L., Tao, Y., 2019. A new generation of AI: a review and perspective on machine learning technologies applied to smart energy and electric power systems. Int. J. Energy Res. 43 (6), 1928–1973.

Cline, B., Niculescu, R.S., Huffman, D., Deckel, B., 2017. Predictive maintenance applications for machine learning. In: 2017 Annual Reliability and Maintainability Symposium (RAMS), Orlando, FL, pp. 1–7. https://doi.org/10.1109/RAM.2017.7889679.

Costello, J.J.A., West, G.M., McArthur, S.D.J., 2017. Machine learning model for event-based prognostics in gas circulator condition monitoring. IEEE Trans. Reliab. 66 (4), 1048–1057. https://doi.org/10.1109/TR.2017.2727489.

Cui, Z., Zhang, Z., Yang, Q., and Kennel, R., 2018. Cascaded model predictive control of three-level NPC back-to-back power converter PMSG wind turbine systems. In: 2018 IEEE International Power Electronics and Application Conference and Exposition (PEAC), Shenzhen, pp. 1–6. doi: 10.1109/PEAC.2018.8590232.

Dalal, G., Gilboa, E., Mannor, S., Wehenkel, L., 2019. Chance-constrained outage scheduling using a machine learning proxy. IEEE Trans. Power Syst. 34 (4), 2528–2540. https://doi.org/10.1109/TPWRS.2018.2889237.

De Benedetti, M., Leonardi, F., Messina, F., Santoro, C., Vasilakos, A., 2018. Anomaly detection and predictive maintenance for photovoltaic systems. Neurocomputing 310, 59–68.

Dobbe, R., Sondermeijer, O., Fridovich-Keil, D., Arnold, D., Callaway, D., Tomlin, C., 2020. Toward distributed energy services: decentralizing optimal power flow with machine learning. IEEE Trans. Smart Grid 11 (2), 1296–1306. https://doi.org/10.1109/TSG.2019.2935711.

Duchesne, L., Karangelos, E., Wehenkel, L., 2017. Machine learning of real-time power systems reliability management response. In: 2017 IEEE Manchester PowerTech, Manchester, pp. 1–6, doi: 10.1109/PTC.2017.7980927.

Dutta, N., Umashankar, S., Shankar, V.K.A., Padmanaban, S., Leonowicz, Z., Wheeler, P., 2018. Centrifugal pump cavitation detection using machine learning algorithm technique. In: 2018 IEEE International Conference on Environment and Electrical Engineering and 2018 IEEE Industrial and Commercial Power Systems Europe (EEEIC/I&CPS Europe), Palermo, pp. 1–6. https://doi.org/10.1109/EEEIC.2018.8494594.

Fang, J., Zhang, X., 2019. Research on power system relay protection method based on machine learning algorithm. In: E3S Web of Conferences. International Conference on Building Energy Conservation, Thermal Safety and Environmental Pollution Control (ICBTE 2019), vol. 136.

Foley, A.M., Leahy, P.G., Marvuglia, A., McKeoghab, E.J., 2012. Current methods and advances in forecasting of wind power generation. Renew. Energy 37 (1), 1–8.

Gan, D., Ke, D.P., 2014. Wind power ramp forecasting based on least-square support vector machine. Appl. Mech. Mater. 535, 162–166.

Garg, A., 2018. Designing Reactive Power Control Rules for Smart Inverters Using Machine Learning. (Master's thesis). Virginia Tech.

Gariba, D., Pipaliya, B., 2016. Modelling human behaviour in smart home energy management systems via machine learning techniques. In: 2016 International Automatic Control Conference (CACS), Taichung, pp. 53–58. https://doi.org/10.1109/CACS.2016.7973883.

Ge, Z., Song, Z., Ding, S.X., Huang, B., 2017. Data mining and analytics in the process industry: the role of machine learning. IEEE Access 5, 20590–20616. https://doi.org/10.1109/ACCESS.2017.2756872.

Gohel, H.A., Upadhyay, H., Lagos, L., Cooperc, K., Sanzetenea, A., 2020. Predictive maintenance architecture development for nuclear infrastructure using machine learning. Nucl. Eng. Technol. 52 (7), 1436–1442.

Guha, N., Wang, Z., Wytock, M., Majumdar, A., 2019. Machine learning for AC optimal power flow. arXiv:1910.08842.

Hariharan, A., Karady, G.G., Dickinson, P.E.J., 2018. Application of machine learning algorithm to forecast load and development of a battery control algorithm to optimize PV system performance in Phoenix, Arizona. In: 2018 North American power Symposium (NAPS), Fargo, ND, pp. 1–6. https://doi.org/10.1109/NAPS.2018.8600594.

Hatziargyriou, N., 2001. Machine learning applications to power systems. In: Paliouras, G., Karkaletsis, V., Spyropoulos, C.D. (Eds.), Machine Learning and Its Applications. In: ACAI 1999. Lecture Notes in Computer Science, vol. 2049. Springer, Berlin, Heidelberg.

Henri, G., Lu, N., Carreio, C., 2018. A machine learning approach for real-time battery optimal operation mode prediction and control. In: 2018 IEEE/PES Transmission and Distribution Conference and Exposition (T&D), Denver, CO, pp. 1–9. https://doi.org/10.1109/TDC.2018.8440141.

Hossain, E., Khan, I., Un-Noor, F., Sikander, S.S., Sunny, M.S.H., 2019. Application of big data and machine learning in smart grid, and associated security concerns: a review. IEEE Access 7, 13960–13988. https://doi.org/10.1109/ACCESS.2019.2894819.

Hu, X., Li, S.E., Yang, Y., 2016. Advanced machine learning approach for lithium-ion battery state estimation in electric vehicles. IEEE Trans. Transp. Electr. 2 (2), 140–149. https://doi.org/10.1109/TTE.2015.2512237.

Idowu, S., Åhlund, C., Schelén, O., 2014. Machine learning in district heating system energy optimization. In: 2014 IEEE International Conference on Pervasive Computing and Communication Workshops (PERCOM WORKSHOPS), Budapest, pp. 224–227. https://doi.org/10.1109/PerComW.2014.6815206.

Idowu, S., Saguna, S., Åhlund, C., Schelén, O., 2016. Applied machine learning: forecasting heat load in district heating system. Energ. Buildings 133, 478–488.

Ishizaki, T., Koike, M., Yamaguchi, N., Ueda, A., Bo, T., Imura, J., 2017. A distributed scheme for power profile market clearing under high battery penetration. IFAC-PapersOnLine 50 (1), 165–170.

Jahnke, P., 2015. Machine Learning Approaches for Failure Type Detection and Predictive Maintenance. (Master's thesis). Technische Universität Darmstadt.

James, S.C., Zhang, Y., O'Donncha, F., 2018. A machine learning framework to forecast wave conditions. Coast. Eng. 137, 1–10.

Johansson, C., Bergkvist, M., Geysen, D., De Somer, O., Lavesson, N., Vanhoudt, D., 2017. Operational demand forecasting in district heating systems using ensembles of online machine learning algorithms. In: 15TH International Symposium on District Heating and Cooling (DHC15-2016). Elsevier, pp. 208–216. https://doi.org/10.1016/j.egypro.2017.05.068.

Jurado, S., Nebot, A., Mugica, F., Avellana, N., 2015. Hybrid methodologies for electricity load forecasting: entropy-based feature selection with machine learning and soft computing techniques. Energy 86, 276–291.

Karagiannopoulos, S., Aristidou, P., Hug, G., 2019. Data-driven local control design for active distribution grids using off-line optimal power flow and machine learning techniques. IEEE Trans. Smart Grid 10 (6), 6461–6471. https://doi.org/10.1109/TSG.2019.2905348.

Karapidakis, E.S., 2007. Machine learning for frequency estimation of power systems. Appl. Soft Comput. 7 (1), 105–114.

Karim, M.A., Currie, J., Lie, T.T., 2018. A machine learning based optimized energy dispatching scheme for restoring a hybrid microgrid. Electr. Power Syst. Res. 155, 206–215.

Keerthisinghe, C., Sun, H., Takaguchi, Y., Nikovski, D., Hashimoto, H., 2018. Machine learning based state-space approximate dynamic programming approach for energy and reserve management of power plants. In: 2018 IEEE Innovative Smart Grid Technologies—Asia (ISGT Asia), Singapore, pp. 669–674. https://doi.org/10.1109/ISGT-Asia.2018.8467807.

Khan, S., Latif, A.F., Sohaib, S., 2018. Low-cost real-time non-intrusive appliance identification and controlling through machine learning algorithm. In: 2018 International Symposium on Consumer Technologies (ISCT), St. Petersburg, pp. 32–36. https://doi.org/10.1109/ISCE.2018.8408911.

Kroll, B., Schaffranek, D., Schriegel, S., Niggemann, O., 2014. System modeling based on machine learning for anomaly detection and predictive maintenance in industrial plants. In: Proceedings of the 2014 IEEE Emerging Technology and Factory Automation (ETFA), Barcelona, pp. 1–7. https://doi.org/10.1109/ETFA.2014.7005202.

Latha, P.G., 2018. A machine learning approach for generation scheduling in electricity markets. Int. J. Electr. Eng. Technol. 9 (3), 69–79.

Leahy, K., Hu, R.L., Konstantakopoulos, I.C., Spanos, C.J., Agogino, A.M., 2016. Diagnosing wind turbine faults using machine learning techniques applied to operational data. In: 2016 IEEE International Conference on Prognostics and Health Management (ICPHM), Ottawa, ON, pp. 1–8. https://doi.org/10.1109/ICPHM.2016.7542860.

Lee, S., Lee, W., 2016. Development of a system for predicting solar power generation and detecting defects using machine learning. KIPS Trans. Comput. Commun. Syst. 5 (10), 353–360. https://doi.org/10.3745/KTCCS.2016.5.10.353.

Li, D., Jayaweera, S.K., 2015. Machine-learning aided optimal customer decisions for an interactive smart grid. IEEE Syst. J. 9 (4), 1529–1540. https://doi.org/10.1109/JSYST.2014.2334637.

Liu, Z., Karimi, I.A., 2020. Gas turbine performance prediction via machine learning. Energy. 192.

Ma, Z., Zhang, C., Qian, C., 2019. Review of machine learning in power system. In: 2019 IEEE Innovative Smart Grid Technologies—Asia (ISGT Asia), Chengdu, China, pp. 3401–3406. https://doi.org/10.1109/ISGT-Asia.2019.8881330.

MacDougall, P., Kosek, A.M., Bindner, H., Deconinck, G., 2016. Applying machine learning techniques for forecasting flexibility of virtual power plants. In: 2016 IEEE Electrical Power and Energy Conference (EPEC), Ottawa, ON, pp. 1–6. https://doi.org/10.1109/EPEC.2016.7771738.

Marvuglia, A., Messineo, A., 2012. Monitoring of wind farms' power curves using machine learning techniques. Appl. Energy 98, 574–583.

Mishra, M., Rout, P.K., 2018. Detection and classification of micro-grid faults based on HHT and machine learning techniques. IET Gener Transm Dis 12 (2), 388–397.

Mulongo, J., Atemkeng, M., Ansah-Narh, T., Rockefeller, R., Nguegnang, G.M., Garuti, M.A., 2020. Anomaly detection in power generation plants using machine learning and neural networks. Appl. Artif. Intell. 34 (1), 64–79. https://doi.org/10.1080/08839514.2019.1691839.

Oluwaseun, A.T., Onyeamaechi, A.M., Sunday, O.S., 2019. Optimization of biogas electrical power generation using neuro-fuzzy controller. Trans. Mach. Learn. Artif. Intell. 7 (6), 21–29. https://doi.org/10.14738/tmlai.76.7239.

Pallonetto, F., De Rosa, M., Milano, F., Finnac, D.P., 2019. Demand response algorithms for smart-grid ready residential buildings using machine learning models. Appl. Energy 239, 1265–1282.

Papakonstantinou, N., Proper, S., O'Halloran, B., Tumer, I.Y., 2014. Simulation based machine learning for fault detection in complex systems using the functional failure identification and propagation framework. In: Proceedings of the ASME 2014 International Design Engineering Technical Conferences and Computers and Information in Engineering Conference. Volume 1B: 34th Computers and Information in Engineering Conference, Buffalo, New York, USA, August 17–20, 2014. V01BT02A022. ASME. https://doi.org/10.1115/DETC2014-34628.

Paterakis, N.G., Mocanu, E., Gibescu, M., Stappers, B., van Alst, W., 2017. Deep learning versus traditional machine learning methods for aggregated energy demand prediction. In: 2017 IEEE PES Innovative Smart Grid Technologies Conference Europe (ISGT-Europe), Torino, pp. 1–6. https://doi.org/10.1109/ISGTEurope.2017.8260289.

Perera, K.S., Aung, Z., Woon, W.L., 2014. Machine learning techniques for supporting renewable energy generation and integration: a survey. In: Woon, W., Aung, Z., Madnick, S. (Eds.), Data Analytics for Renewable Energy Integration. In: DARE 2014. Lecture Notes in Computer Science, vol. 8817. Springer, Cham.

Peters, M., Ketter, W., Saar-Tsechansky, M., Collins, J., 2013. A reinforcement learning approach to autonomous decision-making in smart electricity markets. Mach. Learn. 92, 5–39.

Prior, T., Wäger, P.A., Stamp, A., Widmer, R., Giurco, D., 2013. Sustainable governance of scarce metals: the case of lithium. Sci. Total Environ. 461–462, 785–791.

Provatas, S., 2014. An Online Machine Learning Algorithm for Heat Load Forecasting in District Heating Systems. (Dissertation). Retrieved from: http://urn.kb.se/resolve?urn=urn:nbn:se:bth-3475.

Severson, K.A., Attia, P.M., Jin, N., et al., 2019. Data-driven prediction of battery cycle life before capacity degradation. Nat. Energy 4, 383–391. https://doi.org/10.1038/s41560-019-0356-8.

Shahriar, M.S., Rahman, M.S., 2015. Urban sensing and smart home energy optimisations: a machine learning approach. In: Proceedings of the 2015 International Workshop on Internet of Things Towards Applications (IoT-App '15). Association for Computing Machinery, New York, NY, USA, pp. 19–22. https://doi.org/10.1145/2820975.2820979.

Shao, Z., Wakil, K., Usak, M., Heidari, M.A., Wang, B., Simoes, R., 2018. Kriging Empirical Mode Decomposition via support vector machine learning technique for autonomous operation diagnosing of CHP in microgrid. Appl. Therm. Eng. 145, 58–70.

Sharma, N., Sharma, P., Irwin, D., Shenoy, P., 2011. Predicting solar generation from weather forecasts using machine learning. In: 2011 IEEE International Conference on Smart Grid Communications (SmartGridComm), Brussels, pp. 528–533. https://doi.org/10.1109/SmartGridComm.2011.6102379.

Sica, F.C., Guimarães, F.G., de Oliveira Duarte, R., Reise, A.J.R., 2015. A cognitive system for fault prognosis in power transformers. Electr. Power Syst. Res. 127, 109–117.

Staudt, P., Träris, Y., Rausch, B., Weinhardt, C., 2018. Predicting redispatch in the German electricity market using information systems based on machine learning. In: Proceedings of the 39th International Conference on Information Systems, San Francisco.

Stetco, A., Dinmohammadi, F., Zhao, X., Robu, V., Flynn, D., Barnes, M., Keane, J., Nenadic, G., 2019. Machine learning methods for wind turbine condition monitoring: a review. Renew. Energy 133, 620–635.

Susto, G.A., Schirru, A., Pampuri, S., McLoone, S., Beghi, A., 2015. Machine learning for predictive maintenance: a multiple classifier approach. IEEE Trans. Ind. Inf. 11 (3), 812–820. https://doi.org/10.1109/TII.2014.2349359.

Tan, P., Xia, J., Zhang, C., Fang, Q., Chen, G., 2016. Modeling and reduction of NOX emissions for a 700 MW coal-fired boiler with the advanced machine learning method. Energy 94, 672–679.

Telford, R.D., Galloway, S., Stephen, B., Elders, I., 2017. Diagnosis of series DC arc faults—a machine learning approach. IEEE Trans. Ind. Inf. 13 (4), 1598–1609. https://doi.org/10.1109/TII.2016.2633335.

Ti, Z., Deng, X.W., Yang, H., 2020. Wake modeling of wind turbines using machine learning. Appl. Energy. 257.

Tokel, H.A., Halaseh, R.A., Alirezaei, G., Mathar, R., 2018. A new approach for machine learning-based fault detection and classification in power systems. In: 2018 IEEE Power & Energy Society Innovative Smart Grid Technologies

Conference (ISGT), Washington, DC, pp. 1–5. https://doi.org/10.1109/ISGT.2018.8403343.

Tüfekci, P., 2014. Prediction of full load electrical power output of a base load operated combined cycle power plant using machine learning methods. Int. J. Electr. Power Energy Syst. 60, 126–140.

Uçar, F., Alçin, Ö.F., Dandil, B., Ata, F., 2016. Machine learning based power quality event classification using wavelet—entropy and basic statistical features. In: 2016 21st International Conference on Methods and Models in Automation and Robotics (MMAR), Miedzyzdroje, pp. 414–419. https://doi.org/10.1109/MMAR.2016.7575171.

Ullah, I., Yang, F., Khan, R., Liu, L., Yang, H., Gao, B., Sun, K., 2017. Predictive maintenance of power substation equipment by infrared thermography using a machine-learning approach. Energies 10, 1987.

Umehara, M., Stein, H.S., Guevarra, D., et al., 2019. Analyzing machine learning models to accelerate generation of fundamental materials insights. npj Comput. Mater. 5, 34. https://doi.org/10.1038/s41524-019-0172-5.

Wang, Z., 2019. Predictive control of superheated steam temperature of molten salt heat storage system. In: Proceedings of the International Conference on Advanced Machine Learning Technologies and Applications (AMLTA2019). Springer, pp. 339–345.

Watanabe, F., Kawaguchi, T., Ishizaki, T., Takenaka, H., Nakajima, T.Y., Imura, J., 2018. Machine learning approach to day-ahead scheduling for multiperiod energy markets under renewable energy generation uncertainty. In: 2018 IEEE Conference on Decision and Control (CDC), Miami Beach, FL, pp. 4020–4025. https://doi.org/10.1109/CDC.2018.8619775.

Wehenkel, L., 1997. Machine learning approaches to power-system security assessment. IEEE Expert 12 (5), 60–72. https://doi.org/10.1109/64.621229.

Wood, A.J., Wollenberg, B.F., Sheblé, G.B., 2014. Power Generation, Operation, and Control. Wiley, Hoboken.

Yang, Z., Zhong, J., Wong, S.F., 2011. Machine learning method with compensation distance technique for gear fault detection. In: 2011 9th World Congress on Intelligent Control and Automation, Taipei, pp. 632–637. https://doi.org/10.1109/WCICA.2011.5970591.

Yousefian, R., Kamalasadan, S., 2017. A review of neural network based machine learning approaches for rotor angle stability control. arXiv:1701.01214.

Zaidi, S.M.A., Chandola, V., Allen, M.R., Sanyal, J., Stewart, R.N., Bhaduri, B.L., McManamay, R.A., 2018. Machine learning for energy-water nexus: challenges and opportunities. Big Earth Data 2 (3), 228–267. https://doi.org/10.1080/20964471.2018.1526057.

Zárate, J., Juárez-Smith, P., Carmona, J., Trujillo, L., de Lara, S., 2020. Modelling the vibration response of a gas turbine using machine learning. Expert. Syst. 37(5).

Zenisek, J., Holzinger, F., Affenzeller, M., 2019. Machine learning based concept drift detection for predictive maintenance. Comput. Ind. Eng. 137.

Zhang, L., Li, Y., 2013. Optimal energy management of wind-battery hybrid power system with two-scale dynamic programming. IEEE Trans. Sustainable Energy 4 (3), 765–773. https://doi.org/10.1109/TSTE.2013.2246875.

Zhang, J.-M., Wang, S.-Q., 2002. A modified neuron model-free controller with PID turning gain for hydroelectric generating units 2, 784–787. https://doi.org/10.1109/ICMLC.2002.1174488.

Zhang, Y.G., Wu, Q.H., Wang, J., Oluwande, G., Matts, D., Zhou, X.X., 2002. Coal mill modeling by machine learning based on onsite measurements. IEEE Trans. Energy Convers. 17 (4), 549–555. https://doi.org/10.1109/TEC.2002.805182.

Zhang, Y., Liu, K., Qin, L., An, X., 2016. Deterministic and probabilistic interval prediction for short-term wind power generation based on variational mode decomposition and machine learning methods. Energy Convers. Manag. 112, 208–219.

Zhou, Y., Zheng, S., 2020. Machine-learning based hybrid demand-side controller for high-rise office buildings with high energy flexibilities. Appl. Energy 262, 114416.

5

Data management from the DCS to the historian and HMI

Jim Crompton

Colorado School of Mines, Golden, CO, United States

Chapter outline

Machine Learning and Data Science in the Power Generation Industry. https://doi.org/10.1016/B978-0-12-819742-4.00005-6
© 2021 Elsevier Inc. All rights reserved.

5.1 Introduction

Power generation or electricity generation is the process of generating electric power from sources of primary energy such as heat (thermal), wind, solar, and chemical energy. Overcoming challenges and improving operations in power generation begins with understanding your data. You need to capture, store, contextualize, and visualize time series information, and make it available to the right people, at the right time, to enable the right decisions. Many operators in the power generation industry are increasingly looking to move toward a data-driven and technology-enabled way of working. Often this is called the Manufacturing 4.0 or Plant 4.0.

Although in principle, this may sound relatively straightforward, the reality is that things have become a lot more complicated since companies deployed their initial plant control systems and operations-based information technology solutions years ago. Organizations face many challenges of scale and performance, data quality, data silos, integration, and cybersecurity. There are also an increasing number of confusing and conflicting messages from within the industry as to how best to support digital operations in order to enable further automation and advanced "Big Data" predictive analysis.

Most companies are trying to evaluate how the new "digital" approach fits and how techniques like machine learning, artificial intelligence, and advanced statistics can improve the performance of plants and operation supply chains. It is not that modeling, simulation, and automation projects are something new to power generation, but the industry has been implementing these techniques and technologies for many years. What is different is that the Plant 4.0 is more instrumented and more connected than ever before. With this degree of digitization, operators can develop data-driven models that combine physics, statistics, and best practice engineering principles in order to gain greater insight on what is going on in the plant, and how they can improve performance of the total asset lifecycle.

Key benefits
- Anticipate problems instead of reacting to them.
- Maximize and focus your maintenance resources.
- Reduce damage to critical equipment.

- Avoid unexpected shutdowns and catastrophic failures.
- Standardize across all critical equipment and across the fleet.

The barriers are usually not the technology. The wise advice is to think about people, process, and technology when starting a Plant 4.0 program. This chapter adds some much-needed consideration around the data foundation as well. The technology is the "bright shiny object" that gets too much attention. The trouble spots are often difficult access to relevant data, uncertain data quality (especially when used outside the infamous functional and asset silos), a resistant organizational culture that still values experience over new statistical models, and often a lack of digital literacy within the organization. All of this is especially true when a company starts adding in data from additional sensors in the plant to the volumes of data they have from transactional and engineering systems.

5.1.1   Convergence of OT and IT

This chapter is about data management. It starts with the generation, transmission, and consumer use measurement system for power. This system-of-system perspective includes the instrumentation and control systems perspective from the ground up, which brings a new data type to traditional analytical projects. This area of Operations Technology, or OT, has developed independently from the Information Technology, or IT, world that companies are probably more familiar with. OT systems have largely been in the domain of electrical engineering, instrumentation, and control systems technology. Recently, this operations world is converging with consumer information technology, whereas previous proprietary systems have become more open. The desired learning outcome of this chapter is to better understand and be able to describe this convergence of OT and IT and understand the vocabulary used.

Why is this important? This convergence brings the physical world of instrumented equipment and meters into the digital world where predictive models can

- Identify and monitor plant and transmission systems performance
- Recognize key ways that the field system can fail
- Develop models to predict those failure points
- Build safe, secure, reliable, and performant systems to optimize all of the elements of the system to increase power generation, while lowering operating costs and minimizing the environmental footprint.

Wearables and mobile devices connect the human operator to the process systems. Sensors measure physical behavior and actuators allow the human to control specific process steps. Data collection and aggregation bridge the gap to the Internet of Things (IoT), or edge computing devices, that bring the process and plant data into the cloud where it can be sent anywhere, processed by large computing resources, remotely monitored, and modeled to create the digital twin of the physical plant system.

The term IT-OT convergence was born out of the process industries. Over the past 40 years, process industry companies have invested billions of dollars in OT systems at the plant level, the large majority of which has been hardware, with software to support, typically including technologies like DCS (distributed control systems), APC (advanced process control), Instrumentation, and Data Historians among many others.

Separately, these same companies have spent billions on enterprise IT systems like ERP (enterprise resource planning), SCM (supply chain management), CRM (customer relationship management), and many more. When viewed from the ISA-95 perspective, the power generation industries are also underinvested in level three technologies. Outside of APM (asset performance management), there is not much IT (or OT) investment in the process industries. MES (manufacturing execution system)/MOM (manufacturing operations management), HMI (human machine interface)/DCS (distributed control system), and other types of similar technology are much more broadly adopted in the batch industries.

In the discrete and batch industries, IT-OT convergence has involved different technologies and has evolved along its own path, often referred to as Factory of the Future. On the IT side, most of the technologies are very similar. One notable addition is PLM (product lifecycle management), which often may be managed separately by an Engineering function.

And, from HMI/DCS and following, there are drastic differences from the process industries. No large integrated automation system, like a DCS with data historian, collecting data and controlling process across potentially 10,000s of i/o at a single digit millisecond rate of response. Instead it is machine, skid, and work-cell centric.

OT in the discrete and batch industries has typically included technologies like:

Batch	Discrete
• Raw material tank management systems • Batch management systems/skids • Packaging machines • Material management systems • Palletizing systems	• CNC machines/machine tools • Stamping machines • SMT machines • Robotics • Material management systems • Inspection/metrology • Assembly lines

In the discrete and batch world, these technologies typically are not integrated at scale and not supported by a data historian. At most, they pass data to enterprise IT systems and are integrated at the line or work-cell level which has some important implications for OT in the batch and discrete industries as compared to the process industries (Littlefield, 2020).

5.1.2 Cybersecurity vulnerabilities

The good news is that the plant and the enterprise are more connected than ever before. The bad news is that as industrial companies and utilities connect more systems through digital transformation, they introduce new risks to their environments.

Top six OT cybersecurity vulnerabilities

1. Overreliance on air gap: Typically, OT systems have been segregated from the company's network IT by an air gap. Oftentimes air gaps are not regularly audited, scanned, or verified to ensure lack of connectivity.
2. Lack of training: Many organizations have not put policies or procedures in place to help employees and vendors consistently utilize good security practices.
3. Poor network segmentation: Organizations often do not effectively utilize the concepts of zones and conduits. By limiting access and egress to specific zones, incidents can be better contained.
4. Poor incident response: Often, documented steps for responding to and containing an incident are limited or worse, nonexistent.
5. Poor password practices: Strong corporate password protection policies are not always carried over to the OT environment.

Operators and administrators may have the same user names for various shifts.
6. Absence of notification or detection: How does an organization know when something is wrong?

5.1.3 A maturity model for OT/IT convergence

One place where one can go to see this convergence taking place is a modern control room. A visitor's eyes will be drawn to all of the screens on the front wall of the operations room. All of the data captured from the field is visualized in a variety of different ways, and process alarms will alert operations staff when equipment or process behavior start to vary from normal conditions, allowing emergency situations to be managed with minimal impact.

These control rooms are getting to the state where they have too much data for the human to comfortably comprehend. So, behind the screens, operators build predictive models (or AI assistants) that help the human operator understand all the big events, and even many of the small events that are going on in the plant. When processes are well known and appropriately instrumented, the control systems can be set to automatic and the human is alerted only when necessary.

The convergence process is not happening overnight, the journey is taking several decades. One way to understand this convergence is through a maturity model. The following describes how OT and IT began to grow toward a common integrated perspective. One key observation is that the industry does not swap out its operations technology as often as you change your smart phone, or as quickly as agile development methods update software and websites. An engineer in the control room will probably come across facilities that are in each of three phases. Dealing with legacy technology and infrastructures is a challenge that all power generation data analytics professionals have to deal with. In one situation the engineers have more data than they need, yet in another situation, they are lacking critical information and are uncertain of the data quality that they have on hand.

Generation One:
- Focus on keeping the plant running.
- Simple data processing: Data In-Data Out to trending reports (Process Book).
- Plant operations centers are local.

Generation Two:
- Data is Contextualized (Asset Framework and Asset Analytics).
- Custom workflows as fit-for-purpose solutions.

- Remote decision support environments begin to lower manning levels.
- Analytics and links to specialized advanced algorithms and modeling are beginning to be used.

Generation Three:
- A Digital Twin concept is developed.
- Predictive capabilities (process alarms, optimization, and event frames) emerge.
- Autonomous (normally unmanned) operations complement field operations.

The basic principle of power generation is the process of changing mechanical energy into electrical energy. Mechanical energy is obtained from heat energy (thermal power plants, gas power plants, etc.), wind (wind mills), water flow (hydropower plants), and solar power plants. The key message from the convergence of OT and IT is about building the bridge between the technology used in the plant, to the processing power, and engineering expertise in the office. The technology is not the hard part of this convergence but it is the organizational change and access to trusted data. The operations groups need to think beyond reacting to what is happening at this moment in the plant, to planning and responding to a "manage-by-exception" mentality where plant surveillance is combined with the predictive power of a digital twin model, and the algorithms that are developed from process and equipment data, and engineering and physics principles.

For their part, the IT group needs to expand their scope beyond the data center and the desktop environment at headquarters, to the plant environment. The digital ecosystem now includes the world of sensors and control systems, of telecommunications all the way to the plant, transmission lines and customer meters focusing on operations reliability, and security of a full lifecycle view of the system.

Let's take a deeper dive into the operational data world and find out just what an operator is dealing with before the data scientists start programming the robots and algorithms to do the heavy lifting.

5.2 Sensor data

Sensor data is the output of a device that detects and responds to some type of input from the physical environment. The output may be used to provide information, or input, to another system or to guide a process. Sensors can be used to detect just about any physical element. Applications include sensors on fuel cells, generators, turbines, and power systems. Here are a few examples of sensors, just to give an idea of the number and diversity of their applications:

- An accelerometer detects changes in gravitational acceleration in a device it is installed in, such as a smartphone or a game controller, to determine acceleration, tilt, and vibration.
- A photosensor detects the presence of visible light, infrared transmission (IR), and/or ultraviolet (UV) energy.
- Lidar, a laser-based method of detection, range finding, and mapping, typically uses a low-power, eye-safe pulsing laser working in conjunction with a camera.
- A Charge-Coupled Device (CCD) stores and displays the data for an image in such a way that each pixel is converted into an electrical charge, the intensity of which is related to a color in the color spectrum.
- Smart grid sensors can provide real-time data about grid conditions, detecting outages, faults and load, and triggering alarms.

Wireless sensor networks combine specialized transducers with a communications infrastructure for monitoring and recording conditions at diverse locations. Commonly monitored parameters include temperature, humidity, pressure, wind direction and speed, illumination intensity, vibration intensity, sound intensity, power-line voltage, chemical concentrations, pollutant levels, and vital body functions.

Sensor data is an integral component of the increasing reality of the Internet of Things (IoT) environment. In the IoT scenario, almost any entity imaginable can be outfitted with a unique identifier (UID) and the capacity to transfer data over a network. Much of the data transmitted is sensor data. The huge volume of data produced and transmitted from sensing devices can provide a lot of information but is often considered the next big data challenge for businesses. To deal with that challenge, sensor data analytics is a growing field of endeavor (Rouse et al., 2015).

According to Talend, a Big Data Software company, sensor data and sensor analytics are poised to become the next Big Thing in information technology, with experts predicting that the volume of sensor data will soon dwarf the amount of data that social media is currently producing. Gathered from cell phones, vehicles, appliances, buildings, meters, machinery, medical equipment, and many other machines, sensor data will likely completely transform the way organizations collect information and process business intelligence.

Working with sensor data is problematic for most organizations today. Most enterprises are not equipped to integrate data from such a wide range of sensor sources—their legacy integration technology simply is not up to the task. On top of infrastructure problems, few organizations have developers with the skills required to manipulate massive sets of sensor data, and training existing developers is extremely costly and time consuming (Pearlman, 2019).

5.2.1 Vibration data

While all machinery vibrates, monitoring excess vibration in rotating equipment is key to early identification of asset issues. Vibration monitoring devices use accelerometers to measure changes in amplitude, frequency, and intensity of forces that damage rotating equipment. Studying vibration measurements allows teams to discover imbalance, looseness, misalignment, or bearing wear in equipment prior to failure.

The problem in managing vibration data is twofold. First, many useful applications of vibration monitoring require data at such short time intervals that the data quantity is huge. This creates a problem for the entire infrastructure from sensor, transmission, storage, analysis. Second, a single measurement is not a number but a vector because we measure a whole spectrum that needs its own visualization and expertise.

A number of vendors offer a wide range of Shock, Vibration, and Acceleration data loggers. A shock data logger or vibration data logger is a measurement instrument that is capable of autonomously recording shocks or vibrations over a defined period of time. Digital data is usually in the form of acceleration and time. The shock and vibration data can be retrieved (or transmitted), viewed, and evaluated after it has been recorded.

Applications for these devices include monitoring valuable shipments during transportation or research and development. Transportation loggers are designed specifically for monitoring valuable shipments and have enough memory to record more than 1000 shock and jolt events as well as recording temperature data. Some models expand on that and also record humidity, pressure, and light. Vibration loggers have sample rates as high as 1600 samples per second and are ideal for testing and research applications. Force and acceleration data is represented as G Force and shown for each axis. Handheld vibration meters are ideal for monitoring equipment and predictive maintenance applications.

Vibration analysis (VA), applied in an industrial or maintenance environment, aims to reduce maintenance costs and equipment downtime by detecting equipment faults. VA is a key component of a condition monitoring (CM) program and is often referred to as predictive maintenance. Most commonly VA is used to detect faults in rotating equipment (Fans, Motors, Pumps, and Gearboxes, etc.) such as Unbalance, Misalignment, rolling element bearing faults, and resonance conditions.

VA can use the units of displacement, velocity, and acceleration displayed as a time waveform (TWF), but most commonly the spectrum is used, derived from a fast Fourier transform of the time waveform. The vibration spectrum provides important frequency information that can pinpoint the faulty component (Crawford, 1992).

5.2.2 There are problems with data from sensors: Data quality challenges

Currently, the most common power plants are thermal power plants which derive mechanical energy from heat energy. To control the process and efficiently produce electricity requires a large amount of instrumentation. Flow, level, and temperature measurement are the key areas of process information that need to be monitored and controlled in thermal power plants. In today's data-driven world, it has become essential for companies using IoT technologies to address the challenges posed by exploding sensor data volumes; however, the sheer scale of data being produced by tens of thousands of sensors on individual machines is outpacing the capabilities of most industrial companies to keep up.

According to a survey by McKinsey, companies are using a fraction of the sensor data they collect. For example, in one case, managers interviewed for the survey said they only used 1% of data generated by their facilities 30,000 sensors when making decisions about maintenance planning. At the same time, McKinsey found serious capability gaps that could limit an enterprise's IoT potential. In particular, many companies in the IoT space are struggling with data extraction, management, and analysis.

Timely, and accurate, data is critical to provide the right information at the right time for business operations to detect anomalies, make predictions, and learn from the past. Without good quality data, companies hurt their bottom line. Faulty operational data can have negative implications throughout a business, hurting performance on a range of activities from plant safety, to product quality, to order fulfillment. Bad data has also been responsible for major production and/or service disruptions in some industries (Moreno, 2017).

Sensor data quality issues have been a long-standing problem in many industries. For utilities who collect sensor data from electric, water, gas, and smart meters, the process for maintaining sensor data quality is called Validation, Estimation, and Editing, or VEE. It is an approach that can be used as a model for other industries, as well when looking to ensure data quality from high volume, high velocity sensor data streams.

Today, business processes and operations are increasingly dependent on data from sensors, but the traditional approach of periodic sampling for inspecting data quality is no longer sufficient. Conditions can change so rapidly that anomalies and deviations may not be detected in time with traditional techniques.

Causes of bad sensor data quality include:

- Environmental conditions—vibration, temperature, pressure, or moisture—that can impact the accuracy of measurements and operations of asset/sensors.
- Misconfigurations, miscalibrations, or other types of malfunctions of asset/sensors.
- Drift where an initially calibrated sensor moves away from its calibration over a long period.
- Different manufacturers and configurations of sensors deliver different measurements.
- Loss of connectivity interrupts the transmission of measurements for processing and analysis.
- Tampering of sensor/device and data in transit, leading to incorrect or missing measurements.
- Loss of accurate time measurement because of use of different clocks, for example.
- Out-of-order, or delayed, data capture and receipt.

5.2.3   Validation, estimation, and editing

The goal of VEE is to detect and correct anomalies in data before it is used for processing, analysis, reporting, or decision-making. As sensor measurement frequencies increase, and are automated, VEE is expected to be performed on a real-time basis in order to support millions of sensor readings per second, from millions of sensors.

The challenge for companies seeking actionable operational and business knowledge from their large-scale sensor installations is that they are not able to keep up with the VEE processes needed to support data analytics systems because of the high volume and velocity of data. This is why so any companies today are using less than 10% of the sensor data they collect (Tomczak, 2019).

5.2.4 Time series data

A time series is a series of data points indexed, listed, or graphed in time order. Most commonly, a time series is a sequence taken at successive equally spaced points in time. Thus it is a sequence of discrete time data. Examples of time series are heights of ocean tides, counts of sunspots, pressure and temperature readings, and the daily closing value of the Dow Jones Industrial Average.

Time series are very frequently plotted via line charts. They are used in statistics, signal processing, pattern recognition, econometrics, mathematical finance, weather forecasting, earthquake prediction, electroencephalography (heart monitoring), control engineering, astronomy, communications engineering, and largely in any domain of applied science and engineering which involves temporal measurements.

Time series analysis comprises methods for analyzing time series data in order to extract meaningful statistics and other characteristics of the data. Time series forecasting is the use of a model to predict future values based on previously observed values. While regression analysis is often employed in such a way as to test theories that the current values of one or more independent time series affect the current value of another time series, this type of analysis of time series is not called "time series analysis," which focuses on comparing values of a single time series, or multiple dependent time series at different points in time. Interrupted time series analysis is the analysis of interventions on a single time series.

Time series data have a natural temporal ordering. This makes time series analysis distinct from cross-sectional studies in which there is no natural ordering of the observations (e.g., explaining people's wages by reference to their respective education levels, where the individuals' data could be entered in any order). Time series analysis is also distinct from spatial data analysis where the observations typically relate to geographical locations. A stochastic model for a time series will generally reflect the fact that observations close together in time will be more closely related than observations further apart. In addition, time series models will often make use of the natural one-way ordering of time so that values for a given period will be expressed as deriving in some way from past values, rather than from future values. Transactional data is recorded in a tabular format with values associated by columns in each row. Real-time data is recorded with only time context (i.e., value and timestamp) (Ullah, 2013).

Time series analysis can be applied to real-valued, continuous data, discrete numeric data, or discrete symbolic data (i.e., sequences of characters, such as letters and words in the English language). Methods for time series analysis may be divided into two classes: frequency-domain methods and time-domain methods. The former includes spectral analysis and wavelet analysis; the latter include auto-correlation and cross-correlation analysis. In the time domain, correlation and analysis can be made in a filter-like manner using scaled correlation, thereby mitigating the need to operate in the frequency domain.

Additionally, time series analysis techniques may be divided into parametric and nonparametric methods. The parametric approaches assume that the underlying stationary stochastic process has a certain structure which can be described using a small number of parameters (for example, using an autoregressive or moving average model). In these approaches, the task is to estimate the parameters of the model that describes the stochastic process. By contrast, nonparametric approaches explicitly estimate the covariance or the spectrum of the process without assuming that the process has any particular structure. Methods of time series analysis may also be divided into linear and nonlinear, and univariate and multivariate (Lin et al., 2003).

5.2.5 How sensor data is transmitted by field networks

The goal of all information technology implementations, regardless of industry, should be to improve productivity. The bottleneck for field and plant data flow until now has been the transfer of real-time data to the control room or engineers' desktop in an accurate, timely, and useful fashion. Engineers typically have seen only a subset of the plant data available (a few gauge pressures and temperature settings). With databases updated only periodically from real-time historians, engineers have lacked sufficient insight into the dynamics of plants or field operations. What's needed, "is an alarm system to inform engineers of under-performing or critical conditions," before it begins to degrade production and the revenue stream. Operations need to move beyond the familiar data management mantra of the "right data to the right person at the right time" and adopt the 21st century goal of "validated data, to the decision maker, before the critical event" (Piovesan et al., n.d.).

As field automation moved from plants, where wired connections are possible, to the field, the evolution of radio transmission networks began. There are countless radio networks monitoring and controlling municipal water, wastewater, drainage, oil and gas production, power grids, automatic meter reading, and much more. These radio technology-based communications systems have ranges from yards to hundreds of miles.

Selecting the right technology, defining the projects requirements, and creating a robust communications network are crucial. Having an understanding of the advantages and disadvantages of each technology can help the engineer, or operator, to make the best business decisions for a system that will have a 10- to 20-year service lifetime and are often mission critical or high importance systems.

Here are just a few examples of field, or plant, to office communications protocols.

The HART Communications Protocol (Highway Addressable Remote Transducer) is an early implementation of Fieldbus, a digital industrial automation protocol. Its most notable advantage is that it can communicate over legacy 4–20 mA analog instrumentation wiring, sharing the pair of wires used by the older system.

Fieldbus is the name of a family of industrial computer network protocols used for real-time distributed control, standardized as IEC 61158. A complex automated industrial system—such as manufacturing assembly line—usually needs a distributed control system—an organized hierarchy of controller systems—to function. In this hierarchy, there is usually a Human-Machine Interface at the top, where an operator can monitor or operate the system. This is typically linked to a middle layer of programmable logic controllers via a nontime-critical communications system. At the bottom of the control chain is the fieldbus that links the PLCs to the components that actually do the work, such as sensors, actuators, electric motors, console lights, switches, valves, and contactors.

Requirements of Fieldbus networks for process automation applications (i.e., flowmeters, pressure transmitters, and other measurement devices and control valves in industries such as hydrocarbon processing and power generation) are different from the requirements of Fieldbus networks found in discrete manufacturing applications, such as automotive manufacturing, where large numbers of discrete sensors are used, including motion sensors, position sensors, and so on. Discrete Fieldbus networks are often referred to as "device networks" (Anderson, 2009).

OPC (OLE for Process Control) was first defined by a number of players in automation together with Microsoft in 1995. Over the following 10 years it became the most used versatile way to communicate in the automation layer in all types of industry. Over the years it has evolved from the start with simple Data access (DA), over Alarm and Events (AE), to the more advanced Historical Data

Access (HDA) to have quite extensive functionality and reach. Though there were always some gaps where it did not cover the needs and requirements from the more advanced control systems, it was out of those needs for model-based data and getting more platform independent that resulted in the creation of the OPC UA standard (Marffy, 2019).

Building on the success of OPC Classic, OPC UA was designed to enhance and surpass the capabilities of the OPC Classic specifications. OPC UA is functionally equivalent to OPC Classic, yet capable of much more:

1. Discovery: find the availability of OPC Servers on local PCs and/or networks.
2. Address Space: all data is represented hierarchically (e.g., files and folders) allowing for simple and complex structures to be discovered and utilized by OPC Clients.
3. On-Demand: read and write data/information based on access permissions.
4. Subscriptions: monitor data/information and report by exception when values change based on a client's criteria.
5. Events: notify important information based on client's criteria.
6. Methods: clients can execute programs, etc. based on methods defined on the server.

Integration between OPC UA products and OPC Classic products is easily accomplished with COM/Proxy wrappers (Unified Architecture, 2019).

The most significant difference between classical OPC and OPC UA (OPC Unified Architecture) is that it does not rely on OLE or DCOM technology from Microsoft that makes implementation possible on any platform (i.e., Apple, Linux [JAVA], or Windows). The other very important part of UA is the possibility to use structures or models. This means that the data tags, or points, can be grouped and be given context, making governance and maintenance much easier. These models can be identified at runtime, which makes it possible for a client to explore connection possible by asking the server.

The information modeling is very modern in OPC UA. These models can be defined by manufactures, or protocols like BACNet, but it can also contain more of a MESH structure where very complex relations and connections between points and nodes can be defined. The possibility also exists to have data structures so that certain data always is grouped and handled as one piece. This is important in many applications where you want to be sure that the dataset is taken at the same time.

OPC UA, as said before, is built to be platform independent and the communication is built into layers on top of the standard TCP/IP stack. Above the standard transport layers there are two layers, one that handles the session, and one to establish a secure channel between the client and server. The transport layer is made up of TCP/IP and on top of that SSL, HTTP, or HTTPS. The Communication layer secures the communication channel, not just that the data is corrupted but also it secures the authentication so that the end points cannot be infiltrated and changed. This is based on X.509 certificates that have three parts to it and the first peer-to-peer trust needs to be manually done, but after that the rest is taken care of securely.

With approximately 30 million HART devices installed and in service worldwide, HART technology is the most widely used field communication protocol for intelligent process instrumentation. With the additional capability of wireless communication, the legacy of benefits this powerful technology provides continues to deliver the operational insight users need to remain competitive.

Even though millions of HART devices are installed worldwide, in most cases the valuable information they can provide is stranded in the devices. An estimated 85% of all installed HART devices are not being accessed to deliver device diagnostics information, with only the Process Variable data communicated via the 4–20 mA analog signal. This is often due to the cost and the difficulty of accessing the HART information.

5.2.6 Which protocol is best?

The right answer is "it depends on the application and what is already installed." An optimized solution will probably use more than one communication type, for example, new "analog" installations benefit greatly from FOUNDATION Fieldbus, while new "discrete" installations can benefit from PROFIBUS, PROFINET, DeviceNet, and Ethernet/IP. Electrical Integration depends on the equipment supported protocol, but IEC 61850, Ethernet/IP, and PROFIBUS could all prove useful. The most common view is to look at the current installed base, HART, PROFIBUS, DeviceNet, MODBUS, or others. Why replace when one can integrate?

For the decision maker, it comes down to deciding between a large capital investment and no monthly service fees, OR a smaller capital investment with monthly service fees. If the client's remote assets are well within the service providers' coverage area, they have high speed or TCP/IP data requirements, or their remote sites are there for a short term (lasting no more than 1 or 2 years) all of these factors contribute to making Cellular communications a more attractive option.

If the assets are extremely remote (hundreds or thousands of miles away from public network access) and/or they have TCP/IP

LAN speed requirements, then Satellite communications is obviously the only solution. Often it is more typical for companies to have a mixture of solutions, using Point-Multipoint Private Radio communications where there is a high density of remote sites served by the one Master site, using Cellular for sites for from Private radio clusters, and/or satellite for the very remote sites.

Another difference between Private Radio networks and Cellular or Satellite networks is expansion capabilities. With Private Radio systems, if distances between communicating sites are too great for a single hop, then repeaters can always be installed to further the range. This certainly adds to the overall project cost, but it is at least an option if required.

With Cellular or Satellite, one is leveraging the vast existing infrastructure of the Service Provider. The advantage of Cellular or Satellite is that the client has access to very wide coverage but the disadvantage is that if there is a site that happens to be outside of the Service Providers coverage, there is no way for the client to increase the range; only the Service Provider can do that (Bentek Systems).

5.3 How control systems manage data

Supervisory Control and Data Acquisition (SCADA) is a control system architecture that uses computers, networked data communications, and graphical user interfaces for high-level process supervisory management, but uses other peripheral devices, such as programmable logic controllers and discrete PID controllers to interface to the process plant operations or critical equipment. The operator interfaces which enable monitoring and the issuing of process commands, such as controller set point changes, are handled through the SCADA supervisory computer system. The real-time control logic, or controller calculations, are performed by networked modules which connects to the field sensors and actuators. The main purpose of a SCADA system is to collect data from sensors on well heads, tanks, and critical equipment (i.e., temperature, pressure, vibrations, flowrates, fill levels and provide a view to the human operator the condition of field operations).

5.3.1 Historians and information servers as a data source

A Data Historian, also known as a Process Historian or Operational Historian, is a software program that records and retrieves production and process data by time. It stores the information in a time series database that can efficiently store data with minimal disk space and fast retrieval. Time series information is often displayed in a trend or as tabular data over a time range (i.e., the last day, last 8 h, last year).

A historian will record data over time from one or more locations for the user to analyze. Whether one chooses to analyze a valve, tank level, fan temperature, or even a network bandwidth, the user can evaluate its operation, efficiency, profitability, and setbacks of production. It can record integers (whole numbers), real numbers (floating point with a fraction), bits (on or off), strings (ex. product name), or a selected item from a finite list of values (ex. Off, Low, High).

Some examples of what might be recorded in a data historian include

- Analog Readings: temperature, pressure, flowrates, levels, weights, CPU temperature, mixer speed, fan speed.
- Digital Readings: valves, limit switches, motors on/off, discrete level sensors.
- Product Info: product ID, batch ID, material ID, raw material lot ID.
- Quality Info: process and product limits, custom limits.
- Alarm Info: out of limits signals, return to normal signals.
- Aggregate Data: average, standard deviation, moving average.

A Data Historian could be applied independently in one or more areas, but can be more valuable when applied across an entire facility, many facilities in a department, and across departments within an organization. An operator can discover that a production problem's root cause is insufficient power supply to the production equipment, or they could discover the two similar units produce significantly different results over time.

A Data Historian is not designed to efficiently handle relational data, and relational databases are not designed to handle time series data. If an operator can have a software package that offers both with the ability to integrate data from each different storage type, then they have a much more powerful solution.

There are many uses for a Data Historian in different industries:

- Manufacturing site to record instrument readings.
- Process (ex. flowrate, valve position, vessel level, temperature, pressure).
- Production status (ex. machine up/down, downtime reason tracking).
- Performance monitoring (ex. units/hour, machine utilization vs. machine capacity, scheduled vs. unscheduled outages).

- Product genealogy (ex. start/end times, material consumption quantity, lot number tracking, product set points, and actual values).
- Quality control (ex. quality readings inline or offline in a lab for compliance to specifications).
- Manufacturing costing (ex. machine and material costs assignable to a production).
- Utilities (ex. Coal, Hydro, Nuclear, and Wind power plants, transmission, and distribution).
- Data center to record device performance about the server environment (ex. resource utilization, temperatures, fan speeds), the network infrastructure (ex. router throughput, port status, bandwidth accounting), and applications (ex. health, execution statistics, resource consumption).
- Heavy equipment monitoring (ex. recording of run hours, instrument, and equipment readings for predictive maintenance).
- Environmental monitoring (ex. weather, sea level, atmospheric conditions, ground water contamination).

Information collected within a facility can come from many different types of sources including

- PLCs (Programmable Logic Controllers) that control a finite part of the process (ex. one machine or one processing unit).
- DCS (Distributed Control System) that could control an entire facility.
- Proprietary Instrument Interface (ex. Intelligent Electronic Devices): data delivered directly from an instrument instead of a control system (ex. Weighing system, clean-in-place skid).
- Lab Instrument (ex. Spectrophotometer, TOC Analyzer, Resonance Mass Measurement).
- Manual Data Entry (ex. an operator periodically walks the production line and records readings off manual gauges) (Winslow, 2019).

One of the big issues of historians and all such data management systems is the storage problem. The modern capability of field instrumentations means lots of measurements are recorded. The "Big Data" volume needs lots of storage but traditionally this used to be expensive. With modern storage devices this is no longer the case. But we often find a few legacy rules in place for existing historians that aim to save storage at the expense of data quality/quantity. For example, many existing historians utilize a compression factor to reduce data volumes. Other systems are set up to only keep a set amount of data (say 1 month) and as new data is captured by the historian, older data are rolled off. This is one of the biggest hurdles for big data projects to overcome.

This point leads to the natural question: When ought data points (in a time series) to get stored? Based on time, i.e., after X seconds? When they deviate by X units from the prior point? Something else? In such rules, what is X and who decides this and where is it documented, who decided, and why? How valuable is a full history of operations data and when should we delete? The traditional constraint of economics has become less of an issue but the operations reality of "will anyone actually go back and use historical data records" is still relevant.

5.4 Data visualization of time series data—HMI

Human-machine Interface (HMI) is a component of certain devices that are capable of handling human-machine interactions. The interface consists of hardware and software that allow user inputs to be translated as signals for machines that, in turn, provide the required result to the user. Human-machine interface technology has been used in different industries, such as electronics, entertainment, military, medical, etc. Human-machine interfaces help in integrating humans into complex technological systems. Human-machine interface is also known as man-machine interface (MMI), computer-human interface, or human-computer interface (Techopedia, 2019).

In an HMI system, the interactions are basically of two types (i.e., human to machine and machine to human). Since HMI technology is ubiquitous, the interfaces involved can include motion sensors, keyboards and similar peripheral devices, speech-recognition interfaces, and any other interaction in which information is exchanged using sight, sound, heat, and other cognitive and physical modes are also considered to be part of HMIs.

Although considered as a standalone technological area, HMI technology can be used as an adapter for other technologies. The basis of building HMIs largely depends on the understanding of human physical, behavioral, and mental capabilities. In other words, ergonomics forms the principles behind HMIs. Apart from enhancing the user experience and efficiency, HMIs can provide unique opportunities for applications, learning, and recreation. In fact, HMI helps in the rapid acquisition of skills for users. A good HMI is able to provide realistic and natural interactions with external devices.

The advantages provided by incorporating HMIs include error reduction, increased system and user efficiency, improved reliability and maintainability, increased user acceptance and user comfort, reduction in training and skill requirements, reduction in physical or mental stress for users, reduction in task saturation, and increased economy of production and productivity, etc.

Touchscreens and membrane switches can be considered as examples of HMIs. HMI technology is also widely used in virtual and flat displays, pattern recognition, Internet and personal computer access, data input for electronic devices, and information fusion. Professional bodies like GEIA (Global Emissions InitiAtive) and ISO (International Organization for Standardization) provide standards and guidelines applicable for human-machine interface technology.

5.4.1 Asset performance management systems

Asset performance management should be part of any digital transformation strategy as it is a holistic solution that focuses on asset health. The goal of APM is to identify the leading indicators of equipment problems and then fix them before equipment is damaged, minimizing unplanned down time and unexpected costs.

The objective is to move from the traditional monitoring and surveillance processes where the operator wants to know what happened (descriptive analytics) and why it happened (diagnostic analytics) to a new state where the operator can predict what will happen under certain operating conditions (predictive analytics). It may even be possible in the future to move toward prescriptive analytics where the operator can "tune" the processing system to produce optimum results most of the time (how can we make only good things happen). This end state is sometimes called Asset Performance Management.

Asset Performance Management (APM) is an approach to managing the optimal deployment of assets to maximize profitability and predictability in product supply, focusing on real margin contribution by asset by product code. Rather than looking at an asset on the basis of market value or depreciated value, companies can see how the asset is contributing to their profitability by looking at how individual assets are performing—whether inventory or Plant, Property, and Equipment (PP&E)—and developing a vision of how they want to allocate resources to assets in the future. APM is not necessarily purely financial or even operational, but it will cross functional lines. It combines best-of-breed enterprise asset management (EAM) software with real-time information from production and the power of cross functional data analysis and advanced analytics. More broadly, it looks at the whole lifecycle of an asset, enabling organizations to make decisions that optimize not just their assets, but their operational and financial results as well.

Miklovic (2015)

One of the basic steps of asset performance management is process control and alarm management. Alarm management is the application of human factors (or "ergonomics") along with instrumentation engineering and systems thinking to manage the design of an alarm system to increase its usability. Most often the major usability problem is that there are too many alarms annunciated in a plant upset, commonly referred to as alarm flood (similar to an interrupt storm), since it is so similar to a flood caused by excessive rainfall input with a basically fixed drainage output capacity.

There can also be other problems with an alarm system, such as poorly designed alarms, improperly set alarm points, ineffective annunciation, unclear alarm messages, etc. Poor alarm management is one of the leading causes of unplanned downtime, contributing to over $20B in lost production every year, and of major industrial incidents, such as the one at the Texas City refinery in March of 2005. Developing good alarm management practices is not a discrete activity, but more of a continuous process (i.e., it is more of a journey than a destination) (Mehta and Reddy, 2015).

An Energy Management Platform is a fully delivered, cloud-based energy management solution that utilizes hardware, software, and professional services to gain device-level visibility and actionable intelligence. An energy management platform empowers customers through real-time insight, analysis, and user-level controls to interact and benefit from the expanding smart-grid infrastructure. Benefits of an EMP include

Reporting
- Inputs include energy usage data, energy generation data, sites and zones data, and occupancy and set point data.
- Outputs include base operational reporting.
- Value includes streamline reporting.

Analyzing
- Inputs include HVAC/refrigeration/lighting, monthly bill, and consulting.
- Outputs include performance analytics, alarm prioritization, and automated FDD.
- Value includes asset management and OPEX reduction.

Optimizing
- Inputs include storage solution, interval metering, incentive offer data, and advanced analytics.
- Outputs include optimization, preventative maintenance, utility incentive management, and advanced demand response.
- Value includes asset optimization and CAPEX design.

5.4.2 Key elements of data management for asset performance management

Process historians are the "system of record" for sensor readings, but there are other data management tools that are important as well. Historians store tag names for identification of where a sensor is located in the operations environment, but an operator needs additional contextual information to place the sensor in the operating or plant system. The contextual data, sometimes called master data, is often found in an Asset Registry.

The Asset Registry is an editor subsystem which gathers information about unloaded assets asynchronously as the editor loads. This information is stored in memory so the editor can create lists of assets without loading them. Asset registers are typically used to help business owners keep track of all their fixed assets and the details surrounding them. It assists in tracking the correct value of the assets, which can be useful for tax purposes, as well as for managing and controlling the assets.

There are different ways of building the data model for this kind of information: taxonomy and ontology.

Data taxonomy in this case means an organized system of describing data and information. The information exists. The challenge is to help executives, analysts, sales managers, and support staff find and use the right information both efficiently and effectively. Many enterprises extract value from the business information they accumulate by organizing the data logically and consistently into categories and subcategories; therefore, creating a taxonomy. When information is structured and indexed in a taxonomy, user can find what they need by working down to more specific categories, up to a more inclusive topic, or sideways to related topics (Walli, 2014).

In both computer science and information science, an ontology is a data model that represents a set of concepts within a domain, and the relationships between those concepts. It is used to reason about the properties of that domain and may be used to define the domain. In computer science and information science, an ontology encompasses a representation, formal naming and definition of the categories, properties and relations between the concepts, data and entities that substantiate one, many or all domains of discourse. More simply, an ontology is a way of showing the properties of a subject area and how they are related, by defining a set of concepts and categories that represent the subject.

5.5 Data management for equipment and facilities

In addition to the sensor measurements and control system readings, there are many other important data records available; however, many of these data types are in the form of drawings, permits, documents, inspection records, certificates, P&ID (piping and instrumentation diagram) drawings, process diagrams, etc. A data management system designed for documents is another type of data management technology that is needed.

5.5.1 Document management systems

Document management systems are essentially electronic filing cabinets an organization can use as a foundation for organizing all digital and paper documents. Any hard copies of documents can simply be uploaded directly into the document management system with a scanner. Oftentimes, document management systems allow users to enter metadata and tags that can be used to organize all stored files.

Most document management software has a built-in search engine, which allows users to quickly navigate even the most expansive document libraries to access the appropriate file. Storing sensitive documents as well? Not to worry! Most document management systems have permission settings, ensuring only the appropriate personnel can access privileged information.

These are some of the most important document management features:

1. Storage of various document types, including word processing files, emails, PDFs, and spreadsheets.
2. Keyword search.
3. Permissioned access to certain documents.
4. Monitoring tools to see which users are accessing which documents.
5. Versioning tools that track edits to documents and recover old versions.
6. Controls regulating when outdated documents can be deleted.

7. Mobile device support for accessing, editing, and sharing documents (Uzialko, 2019).

5.5.2   Simulators, process modeling, and operating training systems

Now that you have a basic understanding of the data involved from field instrumentation and process control systems, let's take a look at some of the applications that the sensor data is input to starting with process simulation, or training simulation systems.

Process simulation is a model-based representation of chemical, physical, biological, and other technical processes and unit operations in software. Basic prerequisites are a thorough knowledge of chemical and physical properties of pure components and mixtures, of reactions, and of mathematical models which, in combination, allow the calculation of a process in computers.

Process simulation software describes processes in flow diagrams where unit operations are positioned and connected by product streams. The software has to solve the mass and energy balance to find a stable operating point. The goal of a process simulation is to find optimal conditions for an examined process. This is essentially an optimization problem which has to be solved in an iterative process.

Process simulation always uses models which introduce approximations and assumptions, but allow the description of a property over a wide range of temperatures and pressures, which might not be covered by real data. Models also allow interpolation and extrapolation, within certain limits, and enable the search for conditions outside the range of known properties (Rhodes, 1996).

5.6   How to get data out of the field/plant and to your analytics platform

5.6.1   Data visualization

One area of rapid technology development in this area is the challenge of visualization of time series data. Getting beyond the alarm stage and looking at the data historically to recognize patterns of equipment or process failure and building predictive models to help operators take action before catastrophic failure occur.

5.6.2 From historians to a data infrastructure

While traditional process historians have served an important purpose for control systems for many decades, the limitations of historians are starting to become a barrier for providing data to advanced analytics. The limitations include

- The data is "tag based" creating integration problems with enterprise data access.
- One-off data feeds to specific solutions promote data silos.
- This architecture is difficult to scale and requires expensive maintenance.
- Solutions depend on analyst bringing data together to develop solutions.
- There is limited use of automation to speed up data processing and data exchange.

One leading vendor is this space defines the data infrastructure as a "system of insight" versus the traditional "system of data." The new architecture includes the traditional process historian but adds real-time operational data framework capabilities such as an asset framework to provide asset-based hierarchy context and a library of appropriate data templates to help organize diverse inputs from sensor to meta data streams.

5.6.3 Data analytics

The end goal is to make the operations-oriented sensor, time series data available for

1. Monitoring and Surveillance.
2. Addition of IoT devices, edge, and cloud computing.
3. Alarm Detection (process alarms more than just static set points).
4. Automation of routine tasks.
5. Mobility solutions and advanced visualizations.
6. Integrated Operations.
7. Simulation Modeling.
8. "Big Data" and Data-Driven predictive, "digital twin" models to drive manage-by-exception processes.

These objectives have been coming in stages. The evolution of industrial analytics can be viewed in these three stages of development and maturity.

5.6.4   Three historical stages of industrial analytics

The first stage was defined by employees walking around with pencil and paper reading gauges on various assets in the plant; over time walking around was replaced with wired connections to sensors which used a pen in the control room to write sensor levels on a roll of paper, the "strip chart."

The second stage came with the digitization of sensors and the electronic collection of data, presented to operators on monitors and then stored in historians. Data was viewed with trending applications that started out as electronic versions of strip charts. A connector imported historian data into a spreadsheet and enabled engineers to do calculations, data cleansing, etc. The history of the spreadsheet and historian is closely intertwined—both were invented in the mid-1980s and used in conjunction ever since. Every historian includes at least two application features: a trending application and an Excel connector.

The third stage is currently emerging and assumes changes to the software and architecture for data storage and analytics, tapping advances such as big data and cognitive computing, cloud computing, wireless connectivity, and other innovations.

5.6.5   Where is data analytics headed?

The energy sector collects large amounts of data on a continuous basis. With the applications of sensors, wireless transmission, network communication, and cloud computing technologies, the amount of data being collected on both the supply and demand side of the coin is quite staggering. With the advent of the SMART Grid, this amount of data is only expected to increase. A perfect example of this is how much data 1 million SMART meters collect every 15 min over a year. Two thousand nine hundred and twenty terabytes to be exact and that's only 1 million households or businesses!

One can start expanding that outwards to include other intelligent devices. Examples of these include sensors and thermostats which are used throughout the whole process of power generation, transmission, distribution, and substations. Then the volumes of data collected increases exponentially. Data is only valuable if it's used so the challenge up to now has been to how can businesses and utilities use this "Big Data" in an efficient manner. What value can be derived from all this data?

In the utility industry there are four types of big data sources in utilities: SMART meters, grid equipment, third-party data (off-grid datasets), and asset management data. Utilities are using big data to improve operational efficiencies, drive down costs, and reduce carbon emissions. "Big Data" analytics is also playing a major role in energy management on the demand side. We'll also look at achievements in this space too (Lapping, 2018).

Advanced analytics refers to the application of statistics and other mathematical tools to business data in order to assess and improve practices. In manufacturing, operations managers can use advanced analytics to take a deep dive into historical process data, identify patterns and relationships among discrete process steps and inputs, and then optimize the factors that prove to have the greatest effect on yield. Many global manufacturers in a range of industries and geographies now have an abundance of real-time shop-floor data and the capability to conduct such sophisticated statistical assessments. They are taking previously isolated datasets, aggregating them, and analyzing them to reveal important insights.

Automation increases productivity and helps engineers spend more time on translating data into meaningful information and valuable decisions. Real-time analytics for equipment and operations is providing significant bottom-line value. Multiple predictive analytics are underway to predict compressors and valves failure. The objective is to reduce cost and increase uptime, i.e., compressors trips/failure is one of the top bad actors. Moving to condition-based maintenance will help reduce OPEX and deferment.

Valves example: Valves maintenance is mostly time based; if you are too late, the operator ends up with unscheduled deferment and health, safety, and environmental risks. If your maintenance routine is too early, the operator ends up with scheduled deferment and unnecessary costs.

5.7 Conclusion: Do you know if your data is correct and what do you plan to do with it?

One of the key touch points between factory of the future and the rest of the digitized enterprise is data. Factory of the Future is a major generator and consumer of data to and from many sources and much factory data will be used in the enterprise as businesses are transformed. Defining a data architecture that everyone can work with is a key early stage task in any Industrial Transformation program. A good example is "edge storage" that is a level of storage somewhere between the pure cloud and smart devices on the shop floor. An IT professional may see edge as everything in the plant while a pump maker might see the edge at the interface to his pumps. Everything above the edge can migrate to the cloud but edge computing is required in the plant. Getting an understanding of what that means is critical for a proper understanding between the FoF and IX technical leads and users.

No one role from IT and OT can hope to be able to define a complete data architecture as the scope of systems that will communicate is enormous. What is required for the IX team and its FoF subsidiary is to provide an environment for the changing workforce (Hughes, 2020).

In conclusion, the big question is what is the value of all the effort to collected good operational data for at the moment? Operators have all this equipment and the data is in a database and they have built an HMI or dashboard for surveillance and monitoring. So far so good, but the state-of-the-art use case for this entire Herculean effort is often only to draw timeline diagrams, sit in a conference room, and wait for alarms. Proper analysis is not usually done in most cases and applications such as predictive maintenance, or vibration analysis, or emissions monitoring are the low-hanging fruit available to operators and equipment manufacturers, because all the hard work was done establishing a firm data foundation.

Are there dark clouds on the horizon, holding your data hostage? As organizations look to adopt the new wave of coming technologies (like automation, artificial intelligence, and the Internet of Things), their success in doing so and their ability to differentiate themselves in those spaces will be dependent upon their ability to get operational data management right and find ways to make analysis actionable. This will become increasingly important as connected devices and sensors proliferate, causing an exponential growth in data and a commensurate growth in opportunity to exploit the data.

Those that position their organizations to manage data correctly analyze it in a timely fashion and understand its inherent value will have the advantage. In fact, leaders could pull so far in front that it will make the market very difficult for slow adopters and new entrants.

References

Anderson, M., 7 January 2009. What Is Fieldbus? RealPars. https://realpars.com/fieldbus/.

Crawford, A., 1992. Simplified Handbook of Vibration Analysis. Computational System. ISBN-13: 978-9994616855.

Hughes, A., 11 February 2020. The Factory of the Future and People. LNS Industrial Transformation Blog.

Lapping, D., 12 March 2018. How Big Data Analytics Is Disrupting the Energy Industry. Damon Lapping. https://www.disruptordaily.com/big-data-analytics-disrupting-energy-industry/.

Lin, J., Keogh, E., Lonardi, S., Chiu, B., 2003. A symbolic representation of time series, with implications for streaming algorithms. In: DMKD '03: Proceedings of the 8th ACM SIGMOD Workshop on Research Issues in Data Mining and Knowledge Discovery.

Littlefield, M., 16 January 2020. Why Your Factory of the Future Initiative Will Likely Fail in 2020. LNS Research. https://blog.lnsresearch.com/why-your-fof-initiatives-will-fail-in-2020?utm_campaign=lns-research.

Marffy, S., 2019. Home. Novotek. https://www.novotek.com/en/solutions/kepware-communication-platform/opc-and-opc-ua-explained.

Mehta, B.R., Reddy, Y.J., 2015. Alarm management. In: Alarm Management—An Overview. ScienceDirect Topics, ScienceDirect. https://www.sciencedirect.com/topics/engineering/alarm-management.

Miklovic, D., 16 December 2015. What Is Asset Performance Management? https://blog.lnsresearch.com/what-is-asset-performance-management.

Moreno, H., 5 June 2017. The importance of data quality—good, bad or ugly. Forbes Mag. https://www.forbes.com/sites/forbesinsights/2017/06/05/the-importance-of-data-quality-good-bad-or-ugly/#6b0b9f9d10c4.

Pearlman, S., 28 January 2019. Sensor data—talend. In: Talend Real-Time Open Source Data Integration Software. https://www.talend.com/resources/l-sensor-data/.

Piovesan, C., Kozman, J., Crow, C., Taylor, C., n.d. An Intelligent Platform to Manage Offshore Assets. http://apoffshore.com/wp-content/themes/vanguard/pdf/iPlatform+Magazine+Article.pdf.

Rhodes, C.L., 1996. The process simulation revolution: thermophysical property needs and concerns. J. Chem. Eng. Data 41, 947–950.

Rouse, M., et al., September 2015. What is sensor data?—definition from WhatIs. com. IoT Agenda. https://internetofthingsagenda.techtarget.com/definition/sensor-data.

Techopedia, 2019. What Is Human-Machine Interface (HMI)?—Definition From Techopedia. Techopedia.com. https://www.techopedia.com/definition/12829/human-machine-interface-hmi.

Tomczak, P., 20 May 2019. How VEE Processes Ensure Data Quality for IIoT Systems and Applications. Kx.https://kx.com/blog/kx-for-sensors-data-validation-estimation-and-editing-for-utilities-and-industrial-iot/.

Ullah, M.I., 27 December 2013. Time series analysis. In: Basic Statistics and Data Analysis. p. 2014. Retrieved January 2.

Unified Architecture, 2019. OPC Foundation. https://opcfoundation.org/about/opc-technologies/opc-ua/.

Uzialko, A.C., 21 February 2019. Choosing a document management system: a guide. Business News Daily. https://www.businessnewsdaily.com/8026-choosing-a-document-management-system.html.

Walli, B., 15 August 2014. Taxonomy 101: the basics and getting started with taxonomies. KMWorld. http://www.kmworld.com/Articles/Editorial/What-Is/Taxonomy-101-The-Basics-and-Getting-Started-with-Taxonomies-98787.aspx.

Winslow, R., 2019. What is a data historian? In: Data Historian Overview. http://www.automatedresults.com/PI/data-historian-overview.aspx.

6

Getting the most across the value chain

Robert Maglalang
Value Chain Optimization at Phillips 66, Houston, TX, United States

6.1 Thinking outside the box

In today's global and fast changing business environment, the urgency with multinational companies to find readily implementable digital solutions has increased significantly as the underlying science yielding operational and business improvements has matured over the past decade. Companies lacking innovative ways to extract incremental value from existing and new business ventures will be left behind.

Manufacturing plants, in general, manage billion dollars' worth of raw and finished products daily across the United States and globally. The quantities of data analyzed in a product life cycle, capturing cost-to-produce, logistics, working capital, and other ancillary costs are massive and difficult to both consolidate and integrate. Many production companies still rely heavily on manual processes to evaluate opportunities and economics.

Adopting new technologies has been slow in many industries, and much more so in the energy sector. The traditional, siloed infrastructures to drive quantitatively based decision making

Machine Learning and Data Science in the Power Generation Industry. https://doi.org/10.1016/B978-0-12-819742-4.00006-8
© 2021 Elsevier Inc. All rights reserved.

are not up to the challenge as the processes in place are simply inefficient, thereby creating business risks with regards to data consistency, accuracy, and completeness. In a production plant setting for example, replacing a paper-based system with intrinsically safe gadgets or storing data in a cloud is not an easy transition due to the sensitivity in the security level and risks of causing disruption in the operations. However, since many foundational systems are near the end of useful life—and with renewed focus on improving productivity and yield while cutting down costs—more energy companies are opening their world to the digital technology era.

The plant is the universe around which the organizational improvement focus revolves—a wise investment of resources as they are the outright cash cows. Emphasis on the plant-level optimization and improvements can help a company realize significant value creation through asset reliability and integrity and increase in energy efficiency and productivity. However, putting all strategic efforts alone in the production margin may not move the needle in terms of dollars, especially for the largest players. Outside of this huge "plant box," the greater portion of the business operates in a very competitive market environment where significant opportunities lie. Decision makers require intelligent models with real-time information not only to be immediately available but also to provide valuable insights.

6.2 Costing a project

When implementing ML methods in the industry, we naturally ask: Is it worth it? Measuring the benefit of the technology and its outcomes is not easy. In fact, not even its cost is easy to measure. Fig. 6.1 presents a common situation in large corporations when a significant new initiative is launched. While it is comical to portray the misunderstandings in this fashion, the dire reality is that they often prevent the value from being generated and, in turn, give the technology a bad name.

The point is that before we can talk about measuring the added value of machine learning, we must be very clear and transparent in our communication what we expect it to deliver. Often, industry managers are unsure about what they expect and need help to fully understand even what they could expect. Many vendors choose to dazzle with vocabulary, acronyms, technologies, and dashboards without ever getting down to properly defining the situation—the last image in Fig. 6.1. As John Dewey said, "a problem well stated is a problem half solved."

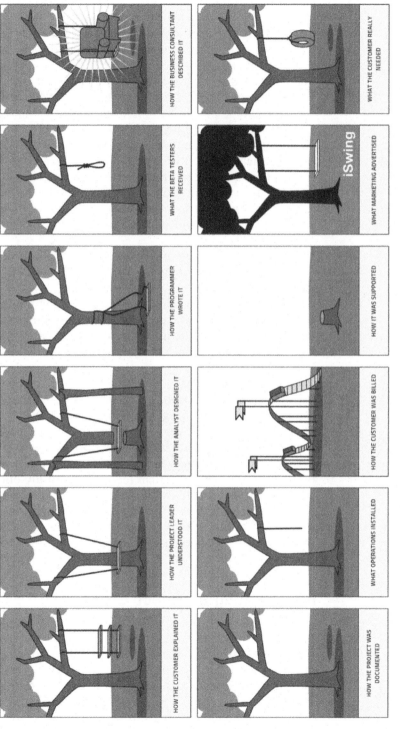

Fig. 6.1 Common problems in communication when implementing a complex project, such as machine learning initiative.

In costing an ML application, consider

- Providing sufficient domain expertise to the machine learners so that they can model the problem in its full practical complexity.
- Involving the end-users to define in what manner they need the answer delivered so that the answer becomes actionable and useful.
- Obtaining good quality, comprehensive, representative, statistically significant, and high-frequency data to enable the data-driven learning to take place.
- Employing expert machine learners to design the algorithms to calculate a precise and accurate answer.
- Spending enough time and effort not just in building the system but testing and debugging it while involving everyone concerned.
- Investing in change management to transition the organization's procedures into being able to utilize the new technology in a practical way.

While business clients focus more on the technological costs of the project, the effort and cost of testing and change management are significant. In fact, most projects that fail, fail because of poorly executed change management. We will discuss this issue later in this chapter.

Most projects that run over the time and financial budget do so because testing and debugging the system takes longer than expected. One way to mitigate this problem is to use the agile methodology of software development—discussed in Chapter 7 of this book—which incorporates feedback into the development cycle. No matter how the project is run however, sufficient attention must be paid to gathering practical feedback from the end-users and adjusting to that feedback. The second most common reason for project failure, after insufficient change management, is a lack of real-world understanding flowing into the analysis. Whenever a vendor claims to be able to solve a problem without detailed and deep domain knowledge, it spells doom from the start.

6.3 Valuing a project

After knowing what it would cost to realize a project, we can think about what it would yield. The reason to think about the benefit after the cost is that the process of costing involved a detailed definition of what the project entails. From experience, projects that were valued prior to a detailed planning were overvalued because of the hype of machine learning that supposedly promises a panacea to all ills in no time at all.

Based on the project definition, what is the kind of benefit being derived? In many cases, there are several benefits. Common dimensions include (1) an increase of revenue, (2) a decrease in cost, (3) a higher production yield, (4) a higher production efficiency, (5) a reduction in staff, (6) a speedup of some kind, (7) consumption of fewer material resources, (8) a reduction in waste or scrap, and (9) a novel business model. Most of these can be measured if we have realistic estimates available.

The last item, a novel business model, is interesting because it is often cited and recognized as very exciting. It is also quite uncertain. Nevertheless, artificial intelligence has enabled numerous new business models that are very profitable and are based essentially in the fast speeds and high complexities that ML models support. Cars without drivers are perhaps the most commonly envisioned transformation in this respect.

In considering the benefits of a project, consider all the stakeholders and their individual benefits as well. They may align with the corporate goals, or not. Also consider the risks if the project does not work out or works less well than imagined. See Fig. 6.2 for an overview of all the factors involved in valuing a project.

A case in point is the famous "pain point" of the customer. Frequently, the issue that generates the most frustration for certain individuals is at the top of their mind and quickly raises the idea

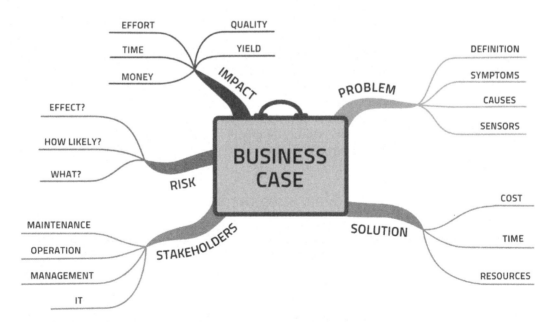

Fig. 6.2 Overview of the drivers in the business case behind a data science project.

that ML could make this issue go away. Enthusiasm is quickly built and a project is envisioned. Vendors are tripping over themselves to fulfill the need and make the customer's pain go away. Finally, a real customer has a real pain point! A sale might be near. Celebrations! But hold your horses. Two separate issues must be examined before too much enthusiasm can be spent on this idea by either the vendor or the people who feel the pain.

First, it must be determined how the benefit is going to be measured. Second, the benefit must then be measured and determined to be large enough. Many situations conspire together to make these two issues work out against doing the project. Let's address both points.

6.3.1   How to measure the benefit

The example for this is predictive maintenance. This is the most discussed topic in the power generation industry where applications of ML are concerned. By the number of articles published and webinars held—by top managers of power generators—on this topic, one would think that this is applied ubiquitously across the industry and world. Far from it. Only a very few companies have deployed it, and even then only in special cases and restricted geographical areas.

One common reason is that maintenance is almost always viewed as an annoying cost item and entirely separated from production. Assessing a maintenance solution by the amount of additional production it enables is obvious. However, most companies do not do their accounting in this way. As a maintenance solution, the maintenance department will have to pay for it and so it must yield a financial benefit within the confines of the maintenance budget. This benefit may be large but the effect on production is almost certainly far greater. Can your organization think broadly enough to consider cross-departmental benefits?

6.3.2   Measuring the benefit

At first glance, measuring the benefit is just an estimate of what the ML project can deliver and putting a financial value to it. However, it is not quite so easy.

All ML methods will produce some false negatives and false positives. In the cases of a regression or forecasting task, the equivalent would be an outlier calculation. No matter what method you employ, it will sometimes make a mistake. Hopefully this will be rare but it will occur and you must consider it.

Statistically speaking, the most dangerous events are those that are very rare and cause large damage. It is very difficult to accurately assess the cost of such events because we cannot reliably assess their likelihood or cost. The most famous example is Deepwater Horizon where a simple cause led to a global disaster. Clearly this type of event is rare and very costly in multiple ways. Fear of such events often leads to an operator not implementing, or not rolling out, an ML initiative. It may represent the main obstacle to full adoption of ML by the industry. The perception of risks threatens to undermine the entire value consideration and so it is important that this be dealt with early on in the discussions, see Fig. 6.3.

6.4 The business case

The business case, then is a combination of five items:
1. Definition of the situation, challenge, and desired solution.
2. Cost and time plan for producing or implementing the solution.
3. Benefit assessment of the new situation.
4. Risk analysis of what might go wrong and lower the benefits or increase the costs.
5. Change management plan of how to implement the solution once it has been created.

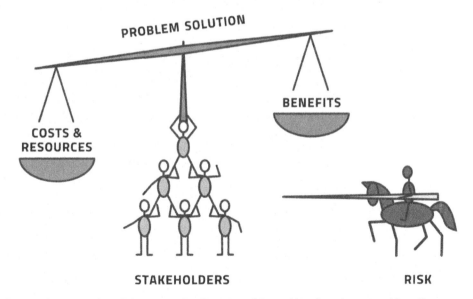

Fig. 6.3 Risk, or the perception of them, may derail any careful consideration of costs and benefits.

If the benefits outweighs the costs—plus or minus the risks—it is worth doing. The assessment is very sensitive to one's point of view as illustrated earlier with the manner in which benefits are calculated.

It is often desired by power generators to conduct a pilot for an ML project in which the ML method is tested out. There are three common dangers in pilot programs. First, pilots often assess only the technology and not the benefits. Second, some pilots try to fully gain and assess the benefits without spending the costs and then obviously fail. Third, most pilots are started using a concrete problem only as a stand-in because the real vision is a vague desire to learn what ML can offer and how it works. This last danger is particularly significant because it means that the organization never intends to put this particular project to productive use.

The business case of ML is similar to all business case arguments. It differs in only two essential ways.

First, the uncertainty of what the solution will look like, which makes the cost and benefit analysis more intransparent.

Second, the expectations and fears of ML in general. Often both are overly inflated by marketing from vendors (who inflate expectations) and cautionary tales by utopians (who are afraid of the robot apocalypse).

The first can be solved easily by some level-headed thinking. The second by talking with some machine learners who are not afraid to say what is realistic and what is squarely within the realm of Hollywood movies.

6.5 Digital platform: Partner, acquire, or build?

In most cases, the software solutions for industrial applications are not readily available. Outside of the plant environment, business-driven analytics and models with accompanying reports and dashboards have not been implemented on a large scale to augment decision making. If vendors have plug and play software that works, the estimated cost is typically a tiny fraction compared to the potential earnings that can be realized if the right solution results in a timely and efficient execution.

To continuously improve and deliver sustainable business performance, companies have put together dedicated resources and formed digital analytics teams within the organization. Relying on internal capabilities is critical to ensure there are clear accountabilities and long-term success. The main players are the data engineers and data scientists combined with subject matter experts and consultants to develop a pipeline of projects around system improvements and automation. With the support from industry consultants, clients evaluate the project trade-offs to set the framework for capital allocation. Also, by going through discussions around effort vs. impact, for instance, they can prioritize high value projects, quick wins, and strategic activities aligned with the overall company objectives and aspirations.

One of the main challenges in launching new applications is timely implementation. Staff needs to understand the complexity of the problems and the right solutions while also figuring out whether to build the software, partner with vendors that have existing or similar type solutions, or acquire the technology outright, if available. There will be trade-offs and conflicting perspectives with these options, and the required implementation timeframe is an important parameter in the decision criteria.

Another consideration is developing a network of excellence with data science and engineers in-house, which is not easy since the roles require programming and IT skills, math and statistics knowledge, and scientific expertise. For power companies, it would take several months or even years to build internal capabilities as these roles are neither their bread and butter nor reside within the organization. Plus, the importance of the full-time internal experts may diminish over time when the technology has been implemented company wide.

A hybrid approach (in-house experts with outside consultants) is more suitable for skill reinforcement and coordination, and to distribute the load evenly. While it is costly to use consultants to provide on-site expertise and support, utilizing them to varying degrees and fostering knowledge sharing could fast track the implementation of the key projects across the enterprise, allowing the company to improve cross functional collaboration and enhance competitive position in the market in a timelier fashion.

6.6 What success looks like

Applying machine learning in the decision process is business critical, most especially when the key factors are unknown, and the quantification of uncertainties and risks can be formulated using existing data. ML has been fully tested in various applications not only to measure opportunities but also to predict and prescribe solutions that enable rapid decisions and deliver profitable growth and enhance returns.

Adoption of ML in the workplace environment is not going to be easy, like any new tool to get the users' buy-in and trust. Change management is fundamental in this process to communicate the scope definition and expectations up front and drive awareness and establish connection points with the impacted stakeholders. With dedicated change management resources, they can support the tool implementation by providing required training, gaining alignment, and creating an open channel for feedback.

Change management is nicely displayed in Fig. 6.4 where we see the evolution of an individual over the course of some change. It begins with anxiety and transitions into happiness that something will change, bringing with it the hope that a pain might go away. But quickly fear settles in as it becomes unclear what impact it will have and might present a bigger change than anticipated. This is followed by guilt at having initiated or participated. At this point, many people chose to stop or become hostile. We must encourage people to see the project as an opportunity here so that they can see it working and see themselves as part of the brave new world.

Guiding people through this process is what defines change management. Without a dedicated change management process, too many people involved will either give up or develop hostility and this leads to project failure. Only with dedicated effort can we bring most, or all, people to the new situation with some enthusiasm and only then will the ML project succeed.

In this effort, it must be recognized that the number of people affected by the new technology is usually quite a bit larger than the group that worked on the project itself. Particularly in the power generation industry, it is the end-users who are expected to use the product on a daily basis that are not involved in the process of making or configuring the product. This alone leads to tensions and misunderstandings. If the end-users are the maintenance engineers who are expected to use predictive maintenance, for example, we may be talking about several hundred individuals who have to transform from reactive to proactive lifestyles at the workplace.

Ultimately, the business solution needs to be simple and streamlined, provide a higher level of accuracy, and give decision makers the right information at their fingertips at the right time. The company significantly benefits financially when existing or new data is examined from a slightly different angle, and derive actionable insights.

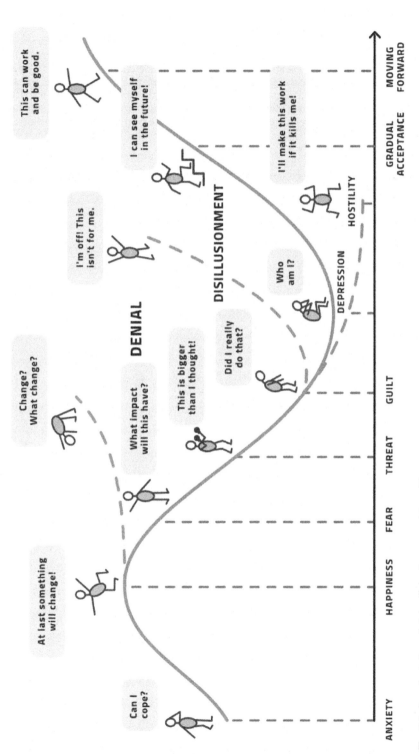

Fig. 6.4 An adaptation of John Fisher's personal transition curve.

Disclaimer

Mr. Robert Maglalang has worked in the oil & gas industry for more than 20+ years in various levels, from refining operations to business development & sales, commercial & marketing, and licensing technology. Opinions and analyses represented here are solely those of the author and do not reflect the views of any persons, companies, or entities referenced within the content. Additional attention has been paid to ensuring compliance with all policies, guidelines, and laws.

7

Project management for a machine learning project

Peter Dabrowski

Digitalization at Wintershall Dea, Hamburg, Germany

Chapter outline

Now that we have a better understanding of machine learning in the context of power generation, we can explore how to best execute these types of projects. A valid question to ask at this point might be: Is the management of projects with a machine learning

Machine Learning and Data Science in the Power Generation Industry. https://doi.org/10.1016/B978-0-12-819742-4.00007-X
© 2021 Elsevier Inc. All rights reserved.

135

component so different that it warrants its own in-depth discussion? The simple answer is "yes," due to the complex and unique nature of machine learning projects, especially when applied to the power industry.

To further answer this question, it is helpful to view machine learning in the context of digitalization. Regardless of chosen technology, cloud computing, smart sensors, augmented reality, or big data analytics, the driver behind digitalization will in many cases be a change in the business model to allow for higher efficiencies and new value adding opportunities. This is not to say the power generation business has not been innovative in the past.

Now, with the application of digitalization, we are leaving the arena of our core business. As we work to marry the traditional technologies with digitalization, what is home turf to Amazon, Google, and Microsoft, we might as well learn from their experiences.

You will notice that this digitalization approach is somewhat different. Buzzwords, such as **Scrum** and **Agile**, will be demystified and translated into our business processes. In the end, you will find that these new processes are yet another set of tools in your toolbox.

7.1   Classical project management in power— A (short) primer

Projects in the power industry are some of the most complex and capital intensive in the world. Major projects, like the building of a new power plant, can cost upwards of US$500 million and make up a significant percentage of a company's expenditures and risk exposure. Therefore focusing on how to best manage processes, dependencies, and uncertainties is imperative.

Traditionally, large projects in engineering design are run sequentially. This sequential approach is phased with milestones and clear deliverables, which are required before the next stage begins. These stages can vary, depending on needs and specifications, but typically all variations contain at least these five basic blocks (Fig. 7.1):

With the workflow cascading toward completion, this approach is often referred to as the Waterfall method. This terminology was applied retroactively with the realization that projects could be run differently (e.g., Agile).

The Waterfall method is one of the oldest management approaches. It is used across many different industries and has clear advantages, including

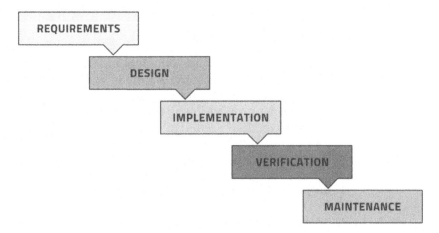

Fig. 7.1 Example workflow of the Waterfall method.

- Clear structure—With a simple framework, the focus is on a limited number of well-defined steps. The basic structure makes it easy to manage, allowing for regular reviews with specific deliverable checks at each phase.
- Focus—The team's attention is usually on one phase of the project at a time. It remains there until the phase is complete.
- Early well-defined end goal—Determining the desired result early in the process is a typical Waterfall feature. Even though the entire process is divided into smaller steps, the focus on the end goal remains the highest priority. With this focus comes the effort to eliminate all risk that deviates from this goal.

Although the Waterfall method is one of the most widely used and respected methodologies, it has come under some criticism as of late. Depending on the size, type, complexity, and amount of uncertainties in your project, it might not be the right fit. Disadvantages of the Waterfall method include

- Difficult change management—The structure that gives it clarity and simplicity also leads to rigidity. As the scope is defined at the very beginning of the process under very rigorous assumptions, unexpected modifications will not be easy to implement and often come with expensive cost implications.
- Excludes client or end-user feedback—Waterfall is a methodology that focuses on optimization of internal processes. If your project is in an industry that heavily relies on customer feedback, this methodology will not be a good fit, as it does not provision these kind of feedback loops.

- Late testing—Verification and testing of the product comes toward the end of a project in the Waterfall framework. This can be risky for projects and have implications on requirements, design, or implementation. Toward a project's end, large modifications and revisions would often result in cost and schedule overruns.

Given its pros and cons and its overall rigid framework, the Waterfall method seems an undesirable management approach for machine learning projects. However, with so many management approaches from which to choose (e.g., Lean, Agile, Kanban, Scrum, Six Sigma, PRINCE2, etc.), how do we know which is best for machine learning projects?

One way to begin to answer this question is to categorize projects based on their complexity. Ralph Douglas Stacey developed the Stacey Matrix to visualize the factors that contribute to a project's complexity in order to select the most suitable approach based on project characteristics (Fig. 7.2).

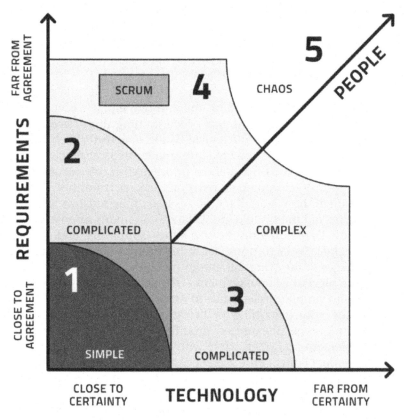

Fig. 7.2 Stacey Matrix with zones of complexity.

On the *y*-axis it measures how close or far members of your team are from agreement on the requirements and objectives of the project. Your team members might have different views on the goals of the project and the needed management style to get there. Your company's governance will influence the level of agreement as well.

Your project's level of certainty depends on the team's understanding of the cause and effect relationships of the underlying technology. A project is close to certain if you can draw on plenty of experience, and you have gone through the planned processes multiple times. Uncertain projects are typically challenged by delivering something that is new and innovative. Under these circumstances experience will be of little help.

Based on these dimensions, we can identify five different areas:

1. Close to agreement/close to certainty—In this zone, we gain information from the past. Based on experience, it is easy to make predictions and outline detailed plans and schedules. Progress is measured and controlled using these detailed plans. Typically, we manage these types of projects using the Waterfall approach.

2. Far from agreement/close to certainty—These projects usually have high certainty around what type of objectives and requirements can be delivered, but less agreement about which objectives are of greatest value. In situations where various stakeholders have different views on added value, the project manager typically has difficulties developing a business case because of the underlying conflicts of interest. Under these circumstances, negotiation skills are particularly important and decision making is often more political than technical. In these instances, the favored management approach is Waterfall or Agile.

3. Close to agreement/far from certainty—Projects with near consensus on the desired goals, but high uncertainty around the underlying technologies to achieve those goals fall into this category. The cause-and-effect linkages are unclear, and assumptions are often being made about the best way forward. The driver is a shared vision by stakeholders that everyone heads toward without specific, agreed-upon plans. Typically, Agile is the approach chosen for these types of projects.

4. The zone of complexity—The low level of agreement and certainty make projects in this zone complex management problems. Traditional management approaches will have difficulties adapting, as they often trigger poor decision-making unless there is sufficient room for high levels of creativity, innovation, and freedom from past constraints to create new solutions.

With adaptability and agility being the key, Scrum and Agile are useful approaches here.

5. Far from agreement/far from certainty—With little certainty and little agreement, we find the area of chaos. The boundary to the complex zone is often referred to as the "Edge of Chaos." Traditional methods of planning, visioning, and negotiating often do not work in this area and result in avoidance. Strategies applied to address these situations are called Kanban or Design Thinking.

Simplifying these different projects down to the degree of available knowledge and characteristics and the responsibilities of a leader highlights the inherent differences.

In Table 7.1, we see how important it is to choose the right process for each project. Although these categories are highly dependent on environment and the team's capabilities (a project that is complicated for an expert can be complex for a beginner), most power projects are typically categorized as complicated. They are characterized by best practices and focus on efficiency. Execution works fantastically well with top-down management and

Table 7.1 Complexity in relation to management style according to the Cynefin framework.

Environment	Characteristics	Leader's job
Chaotic (little is known)	High turbulence	Action to reestablish order
	No clear cause-effect	Prioritize and select work
	True ambiguity	Work pragmatically rather than to perfection, act, sense, respond
Complex (more is unknown than known)	Emerging patterns	Create bounded environment for action
	Cause-effect clear in hindsight	Increase level of communication
		Servant leadership
	Many competing ideas	Generate ideas, probe, sense, respond
Complicated (more is known than unknown)	Experts domain	Utilize experts for insights
	Discoverable cause-effect	Use metrics to gain control
		Sense, analyze, respond
	Processes, standards, manuals	
Simple (everything is known)	Repeating patterns	Use best practices
	Clear cause-effect	Establish patterns and optimize
	Processes, standards, manuals	Command and control, sense, categorize, respond

clean lines of authority for command and control. In these circumstances, the Waterfall method is the best management option.

What about machine learning projects? Application of machine learning and artificial intelligence to modern day problems is an innovative process. When compared with other industries, machine learning in the power industry has only recently found its application. Only governments lag farther behind power even further when comparing industry adoption of digitalization technologies (Source: World Economic Forum).

Machine learning is most effective when applied to complex problems. As outlined earlier, these are projects with many variables and emerging, interdependent interactions. With the interplay between these variables and dependencies being too complicated to predict, upfront planning is useless

As previously stated, Scrum is the most often used management approach to tackle complex projects. Soon, we will dive into the details of applying Scrum to projects, but before doing so, let us highlight the pitfall for managers if this premise is not well understood.

The danger comes in the form of established processes and habits. As mentioned, the majority of power projects are being managed through the Waterfall method. However, a leader tasked with managing a complex machine learning project and using only familiar tools from simple or complicated projects is a recipe for conflict, failure, and misunderstanding.

From Table 7.2, we can see how the characteristics of a complex project, with uncertainty in the process and creative approaches, entail too many competing ideas that rely on the respective skills and competencies of the leader.

For a complex project, a good approach is to rely less on experienced professionals in the specific technical field, but rather, collect various theories and ideas and observe the effect of choices by using an Agile approach. (You can, of course, heavily rely on team members with experience with Agile projects.) The project team must identify, understand, and mitigate risk as new results emerge. This often happens at a rapid pace, requiring a good leader to be an integral team player by enabling the rest of the team and driving cooperation and open communication. This type of leadership is referred to as "servant leadership." In order to arrive at productive solutions for complex projects, teams must approach a problem holistically through probing, sensing, and responding, as opposed to trying to control the situation by insisting on a plan of action.

Table 7.2 The importance of matching the right management style with the respective project type.

Environment	Characteristics	Leader's Job
Chaotic (little is known)	High turbulence No clear cause-effect True ambiguity	Action to re-establish order Prioritize and select work Work pragmatically rather than to perfection, act, sense, respond
Complex (more is unknown than known)	Emerging patterns Cause-effect clear in hindsight Many competing ideas	Create bounded environment for action Increase level of communication Servant leadership Generate ideas, probe, sense, respond
Complicated (more is known than unknown)	Experts domain Discoverable cause-effect Processes, standards, manuals	Utilize experts for insights Use metrics to gain control Sense, analyze, respond
Simple (everything is known)	Repeating patterns Clear cause-effect Processes, standards, manuals	Use best practices Establish patterns and optimize Command and control, sense, categorize, respond

Modifed from http://Scrum.org/PSM.

The potential mismatch between organizational requirements of a successfully managed, complex project and what a typical Waterfall environment provides as outlined in Fig. 7.2 is why we need a different project management approach to machine learning projects. In the next section, we explore the specifics of Agile and Scrum and learn how these approaches are best applied to the world of power.

7.2 Agile—The mindset

Agile, Sprint, Scrum, Product Owner, and Retrospective—welcome to the world of digitalization buzzwords. A world that from the perspective of traditional project management might come across as disorganized—a passing fad, not to be taken seriously. However, by the end of this chapter, you will be able to put meaning to the buzz and understand where and how this approach is best applied.

The term **Scrum** was introduced in 1986 by two Japanese business scholars. They published the article "New New Product Development Game" (that is not a typo) in the Harvard Business Review, describing a new approach to commercial product development that would increase speed and flexibility. Scrum has its roots in manufacturing (e.g., Japanese automotive, photocopier and printer industries) and made its way into the software development industry to establish a new process, as an alternative to the dominant Waterfall method.

In 1995, Jeff Sutherland and Ken Schwaber, two US American software developers, formalized the method in a paper they presented at the OOPSLA conference in Austin, Texas. Their collaboration resulted in the creation of the Scrum Alliance in 2002—a group of pioneers in Agile thinking who came together to evaluate similarities in Agile methods. The result of the group's work was the **Agile Manifesto**, outlining the 12 principles of Agile software development.

In 2009, Ken Schwaber left the Alliance and became the founder of Scrum.org, which oversees the Professional Scrum Certification and publishes an open-source document called The Scrum Guide. The Scrum Guide is today's standard reference guide for Scrum project management.

So, what exactly is Agile and Scrum then? Agile is a general term that describes approaches to product development, focusing on incremental delivery and team collaboration with continual planning and learning. It is a set of principles and values with people, collaboration and interaction at its core and it can also be applied in other fields, such as organizations and training. While Agile provides the mindset, it is Scrum that outlines the concrete framework. It is one of various frameworks under the Agile umbrella, and the one that we will concentrate on for the purpose of managing machine learning projects.

7.3 Scrum—The framework

If you are a fan of the sport of rugby, you certainly have heard of the term Scrum before. This word choice is no coincidence. In rugby, a Scrum (short for scrummage) refers to restarting a play and involves players being densely packed with their heads down, attempting to gain possession of the ball and progress down the field.

The terminology is purposefully chosen by Takeuchi and Nonaka (1986), who already in their original paper pointed toward the analogy: "The traditional sequential or 'relay race' approach to product development [...] may conflict with the goals of maximum speed and flexibility. Instead, a holistic or 'rugby' approach—where a team tries to go the distance as a unit, passing the ball back and forth—may better serve today's competitive requirements" (Fig. 7.3).

Also, Schwaber recognized the similarities between the sport and the management process. The context is the playing field (project environment) and the primary cycle is to move the ball forward (sprint) according to agreed rugby rules (project controls). Further, the game does not end until the environment dictates it (business need, competition, functionality, or timetable). Interestingly, he also acknowledged that rugby evolved from breaking traditional soccer rules, alluding to how Agile is new relative to traditional Waterfall framework.

Fast-forwarding to the present and using *The Scrum Guide* as reference, we note the following definition: "Scum (n): A framework within which people can address complex adaptive problems, while productively and creatively delivering products of the highest possible value."

Scrum is a lightweight framework for enabling business agility. Rather than being a process or technique for building products, it provides the framework within which you can employ various processes and techniques. This framework is based on empirical process control theory, acknowledging that the problem cannot be fully defined or understood up-front. Instead, it focuses on the team's ability to quickly deliver value.

Fig. 7.3 A Scrum during a rugby match. © Luis Escobedo.

Scrum is best understood by investigating its elements. In the following section we will look at the 3-3-3-5-5 framework, which describes the following:

• **3 Pillars of Scrum theory**	– **What is Scrum based on?**
• **3 Roles**	– **Who is involved with which responsibilities?**
• **3 Artifacts**	– **What is being delivered?**
• **5 Events**	– **What happens when?**
• **5 Values of Scrum**	– **Which beliefs motivate actions?**

Let us start with the foundation. Scrum is based on three pillars of process control in which certain assumptions are met (Fig. 7.4):

1. **Transparency**—Important aspects of the process are visible to the entire team, and there is a common understanding of language, including the definition of "done." Everyone knows everything.
2. **Inspection**—Progress and deliverables (artifacts) are often inspected toward their goal to detect undesirable variances. Check your work as you do it.
3. **Adaptation**—If inspection shows that aspects of the process deviate outside acceptable limits resulting in an unacceptable product, the process must be adjusted as soon as possible to minimize further deviation. It is OK to change tactical direction.

Inspection and adaptation are applied systematically throughout the process, as will be evident as we dissect all elements of the Scrum framework.

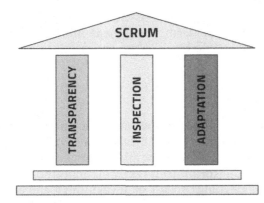

Fig. 7.4 Three pillars of Scrum.

7.3.1 Roles of Scrum

There are three, and only three, roles within the Scrum team: the product owner, the development team, and the Scrum master. Ideally, they should be colocated to foster communication and transparency.

The **product owner** represents the product's stakeholder and voice of the customer. He is responsible to optimize toward adding most value to the product and manages the product backlog. He does not have the responsibilities of a typical manager (i.e., he is not managing a team or a project).

The **development team** typically has three to nine members and focuses on delivering increments of added value to a product during a specific amount of time (a **sprint**). Although members of the team are often referred to as developers, they can often be from various cross functional backgrounds, such as designer, architect, or analyst.

Most importantly though, especially relative to setups in power, is the fact that the team is self-organized. This means that while all work comes through the product owner, all responsibility to deliver its work falls solely on the team itself—no one tells the development team how to do their work. The product owner and the Scrum master are not part of the development team.

The **Scrum master** facilitates the Scrum methodology. He promotes Scrum by making sure everyone understands its practices, rules, and values. By removing possible impediments to the ability of the team to deliver the product goals, the Scrum master acts as the servant leader of the Scrum team. Table 7.3 summarizes the Scrum master's most important services toward stakeholders.

The Scrum master's role differs from that of a project manager because the Scrum master does not have any people management responsibilities. Since the responsibility of managing the development team lies within the team itself, the Scrum master's involvement in the direction setting process is limited. In fact, Scrum does not recognize the role of a project manager.

7.3.2 Events

Scrum has a set of well-defined events to minimize the need for other unnecessary meetings. All events are time boxed, i.e., they have a specified maximum duration. As you will see, all these events (except the sprint itself) provide opportunities to inspect and adapt the ongoing process of development and add to transparency (Fig. 7.5).

Table 7.3 Services of the Scrum master.

Stakeholder	Services of Scrum master
Product owner	• Ensures goals, scope, and product domain are understood by everyone • Helps arranging and prioritizing the product backlog
Development team	• Removes impediments • Coaches team in self-organization • Facilitates Scrum events
Organization	• Help employees and stakeholder understand and enact Scrum

In Scrum, everything revolves around the **sprint**. It is typically between 2 to 4 weeks long (never more) with the goal to produce an increment of useable, potentially shippable product by the end. At the beginning of each sprint (during the **sprint planning**) the team agrees on the scope and defines the goals of the sprint by moving items from the **product backlog** into the **sprint backlog**. During the sprint, no changes are made that endanger the sprint goal, although the scope may be clarified and renegotiated between the product owner and the development team as new insights arise.

Every sprint ends with the **sprint review** and **sprint retrospective**. As soon as one sprint ends, the next begins immediately (Fig. 7.6).

The **sprint planning** meeting marks the beginning of a sprint. A sprint planning, capped at eight hours, focuses on answering two main questions: (1) Which product increment can be delivered in this upcoming sprint (sprint goal)? and (2) How can this chosen work be done?

In practice, the first half of the meeting involves the entire Scrum team (Scrum master, product owner, and development team) suggesting product backlog items practical for the upcoming sprint. Ultimately, the development team selects the specific items for the sprint, as they are best equipped to judge what is achievable.

In the second half of the meeting, the development team breaks down the chosen product backlog items into concrete work tasks, resulting in the sprint backlog. To refine the sprint backlog, the development team might negotiate the selected items with the product owner to agree on the highest priority tasks. In the end, the development team explains to the Scrum master and the product owner how it intends to complete the work.

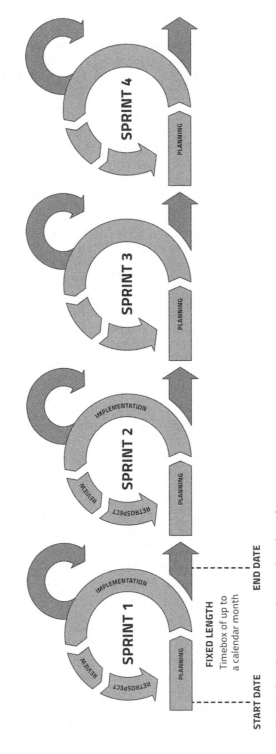

Fig. 7.5 Development cycles are done in sprints.

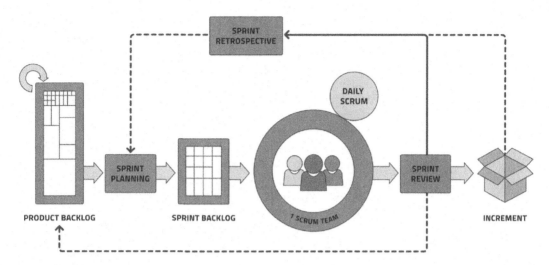

Fig. 7.6 Scrum framework.

The **daily Scrum** is a meeting that is held every day of the Sprint and is time boxed to 15 min. Everyone is welcome, but it is the development team members that contribute. It should always happen at the same time and location and start promptly, even if some team members are not present. The goal of the meeting is for the whole team to answer three questions:

1. What did I do yesterday that contributed to the team reaching the sprint goal?
2. What will I do today that will contribute to the team reaching the sprint goal?
3. Do I see any impediment that could prevent me or the team from reaching the sprint goal?

It is the Scrum master's responsibility to keep the development team focused and to note potential any impediments. The daily Scrum is not to be used for detailed discussions, problem solving, or general status updates. Once the meeting is over, individual members can meet to resolve any open issues in breakout sessions.

The **sprint review** and the sprint retrospective are held at the end of each sprint. They provide the opportunity to assess progress toward the goal and the team and collaboration, respectively. For the purpose of reviewing the sprint progress, the team does the following:

1. Reviews and presents the completed work (e.g., in the form of a demo).
2. Address which items were not "done" (completed).
3. Collaborates on what to work on next.
4. Reviews of timeline, budget, and required capabilities.

Output from the meeting is a revised product backlog and dialog to inform the next sprint planning.

The **sprint retrospective** is an opportunity for the team to reflect and improve. Three questions facilitate this process, including

1. What went well in the last sprint with regards to people, relationships, processes, and tools?
2. What did not go so well?
3. What can be improved to ensure better productivity and collaboration in the next sprint?

The entire team participates in the sprint review and sprint retrospective, which for 4-week sprints, are time boxed to 2 h and 90 min, respectively (and scales proportionally for shorter sprints). For both meetings, the Scrum master is responsible for ensuring all participants to stick to the rules and time limits are honored.

Should it become apparent that a sprint goal is obsolete (due to company strategy, market conditions, technology changes, etc.) during the sprint itself, the product owner may cancel the sprint. However, due to the short duration of most sprints, cancelation is rare.

7.3.3 Artifacts

Another component contributing to transparency in Scrum is the **artifact**. Artifacts provide information on the product, forming a common understanding and allowing for progress on inspection and adaptation. Typical Scrum artifacts include the product backlog, sprint backlog, and product increment.

A **product backlog** is an ordered list of all that is initially known about the product. In addition to foundational knowledge and understandings, it includes a breakdown of work to be done. Business requirements, market conditions, or technology can cause modifications to the backlog, making it an ever-evolving artifact. The listed items can be features, bug fixes, or other nonfunctional requirements. The entire team has access to the product backlog, but the product owner is solely responsible for it. Only the product owner can make changes and set priorities for individual items. Typically, the product owner will gather input and feedback and will be lobbied by various stakeholders; however, it is ultimately his decision on what will be built (and not that of a manager in the organization).

The **sprint backlog** is a set of product backlog items selected for the current sprint. It serves as a forecast of what functionality will be delivered by the end of the sprint and is being modified throughout the sprint to represent progress. Backlog items can be broken up into smaller tasks, which, rather than being assigned, team members will tackle based on priorities and individual skill sets. When the team finalizes the sprint, they analyze and reprioritize the remaining product backlog to have it prepared for the next sprint.

The team's definition of "done" results in a **product increment**, which is a list of all product backlog items that were completed during the sprint. When combined and integrated into all previous increments, the product increment embodies a potentially shippable product that is functional and usable. It is up to the product owner to release it or not (Fig. 7.7).

7.3.4 Values

From the earlier figure, one can tell how all the different elements of Scrum enable a feedback-driven, empirical approach. The three pillars of Scrum—transparency, inspection, and adaptation—require trust and openness. To enable this trust and openness, Scrum relies on following five values:

- **Commitment**—Individual team members personally commit to achieving the goals of the Scrum team.
- **Focus**—Team members focus exclusively on the work defined in the sprint backlog; no other tasks should be done.
- **Courage**—Team members have the courage to address conflict and problems openly to resolve challenges and do the right thing.
- **Respect**—Team members respect each other as individuals who are capable of working with good intent.
- **Openness**—Team members and stakeholders are open and transparent about their work and the challenges they need to overcome obstacles.

These values are essential to successfully using the Scrum method.

7.3.5 How it works

By now you have a rough understanding of Scrum. But if we break it down, what does it mean to execute a project in an Agile way, using the Scrum framework? Let us translate the buzzwords into practice.

Short iteration cycles provided by time-boxed sprints afford high transparency and visibility. With this visibility, and the mandate to self-manage, it is quite easy to maintain adaptability to business and customer needs as well as steer your product in

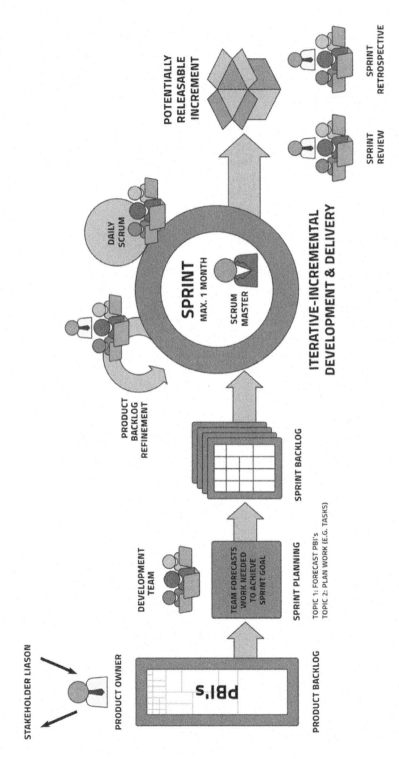

Fig. 7.7 All elements of Scrum during a Sprint.

the right direction. Early product releases at the end of these short iterations help generate value relatively early in time, reducing risk and unwanted cost overruns.

Returning to our comparison of the Scrum framework to traditional project management techniques, Fig. 7.8 illustrates how the differences transpire between the two approaches. Predicting all features of a project with large uncertainty is much more challenging at the beginning of the project, often resulting in surprises at project end when the customers are presented when customers are presented with results when deadlines loom. Agile development allows for better adaptability, lower risk early on, and earlier creation of business value.

The process of creating a car is one commonly used example to illustrate how Agile is applied in the real world and how it differs from big bang deliveries (i.e., build it until it's 100% done, and then deliver it).

Let us assume we have a customer who has ordered a car. Delivery of the first iteration of the product is often misunderstood as delivering an unfinished product. In the case of our example, it would mean the delivery of a tire at iteration 1 (Fig. 7.9). Naturally, the customer will not be happy—after all he has ordered a car, not a tire!

The customer's reaction is not going to be much different at iterations (2, 3), when the product is still a partial car at best.

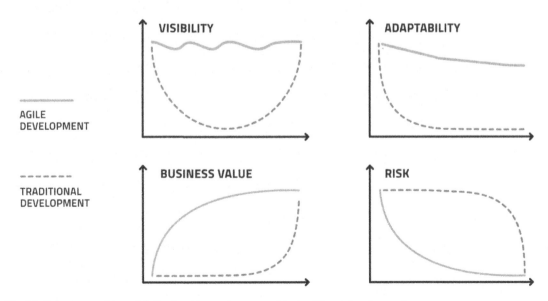

Fig. 7.8 Value proposition of Agile development compared to traditional development.

Fig. 7.9 Building a car—not using proper Agile development.

Although the product is getting closer to its final state, it is not until the final iteration (4) that we have a satisfied customer. And in this example, he is happy because it is what he ordered. A lot of time passes before the customer gets to see any of the product, resulting in the product being based on a lot of assumptions and design flaws.

Now, let us contrast this with the Agile approach. Although the customer ordered a car, we start by focusing on the underlying need he wants to see fulfilled. In this case, conversation with your customer might show that "getting from A to B quickly" is the underlying need, and a car is one of many ways to do that. (Keep in mind that the car is just a metaphor for any kind of customized product development.)

After the first sprint, the team would deliver the most minimalistic thing to get feedback from the customer, e.g., a skateboard. We could call this an **MVP** (**Minimum Viable Product**). Naturally, the customer will likely be unhappy about the result, as it is nowhere near the car, but the goal at this stage is not to make the customer happy, but to get early feedback, learn, and make adaptations for product development down the road.

The next iteration could have a different color (or other changes based on early feedback) and a way of steering, to make it safer to drive around with. The third iteration could be a bicycle, which the customer might start using, resulting in even more valuable. Based on this learning, the 4th sprint might be a motorbike as a deliverable. And at this stage, it may be that the customer is much happier with this product than he thought he would be with the car he originally ordered (Fig. 7.10).

Or, in the final iteration, he sticks to his original idea and gets the car delivered at last, but it's better tailored to the feedback from previous iterations, perhaps resulting in a convertible rather than a coupe. In the end the customer is overjoyed—he got a car and a better one than he originally envisioned!

Fig. 7.10 Building a car—using proper Agile development.

With the underlying question being "what is the cheapest and fastest way we can start learning?" it is possible to reduce risk early and provide the customer with business value that meets or exceeds his exact needs.

7.4 Project execution—From pilot to product

With our newfound understanding of Scrum and Agile, we can now identify opportunities to apply this approach within the context of a power company. Bringing this approach to a business world defined by Waterfall management will be a challenge that is best tackled with the introduction of pilot projects.

A pilot project, by nature of its test setup, gets the mandate of experimentation. It is usually run in parallel to day-to-day operational business and focuses on the validation of a technology or process. In our case, this is the application of our machine learning algorithm. Let us use an example to examine a possible pilot setup and see how the different roles and stakeholders would interact. We will end the chapter outlining how you can introduce Scrum as a possible management framework within your company.

7.4.1 Pilot setup

ABC Power Ltd., a small power company, is interested in exploring machine learning in the context of a predicting gas turbine failures. Under normal circumstances, ABC Power conducts periodic inspections of its gas turbines. In the interim, plant engineers rely on extrapolating from previous test results, which leads to uncertainty and inaccuracies in data estimates.

By incorporating machine learning, advanced analytics, and all available production data (e.g., pressures, temperatures, flows, and vibrations), ABC Power sees an opportunity in using predictive analytics to forecast near-term future failures. This innovation would allow for informed decision-making based on live data and preventative shutdowns.

ABC Power has identified SmartML Ltd. as a potential partner for developing and testing this predictive technology. SmartML has experience working with companies in the power industry, but its main expertise lies in the application of artificial intelligence and machine learning. ABC Power and SmartML agree to cooperate and perform a pilot investigating the effectiveness of applying machine learning to all available data to create a forecast model.

Both companies are aware of the uncertainties, accept them, and agree to run this project in using the Scrum framework, including filling roles of product owner, Scrum master, and development team.

7.4.2 Product owner

A technical representative from ABC Power Ltd. serves as the product owner. As an ABC employee, she possesses the best understanding of the requirements of the product (a gas turbine failure forecaster) and thereby, she is responsible for setting priorities for the backlog items and making sure the team understands the product goal.

However, given that this pilot project is occurring concurrently with existing ABC Power operations, both companies see benefit in identifying a proxy product owner who serves as the go-between for the product owner and the development team. The proxy product owner is part of SmartML and thus, is more available to dedicate time to the project than the product owner, who is also managing ABC Power's day-to-day operations. The product owner and the proxy product owner require a relationship based on trust and open communication.

7.4.3 Development team

In the pilot, SmartML's developers and designers make up the core of the development team. Generally, the composition of the team depends on the size and capabilities of the power company. The larger the power company, the more capabilities of internal capacity it can provide, while smaller companies may only possess the capacity to supply the product owner.

In the case of ABC Power, the more interdependencies that exist between software solutions and the existing infrastructure, of ABC Power the more important it is to involve ABC IT and OT engineers and data architects on the development team. Connecting relevant data sources and maintaining cybersecurity, while simultaneously allowing an external partner access to data, can be challenging and requires prioritization and teamwork.

Depending on the skill set available at ABC Power, it may or may not involve its own staff during the development phase, but certainly will involve staff in the testing and validation process when the team finishes the first iterations of the solution. To this end, there is a notable benefit to having the team colocated. The larger and more complex the project, the more important colocation becomes in terms of facilitating interactions and improving transparency and clarity—especially at the beginning of a project when trust is being established. Remember, that the team is self-organized, so, neither the product owner, her proxy, nor the Scrum master, act as managers or superiors in this setup.

7.4.4 Scrum master

The Scrum master focuses on facilitation, meaning this role can be filled by ABC Power, SmartML, or an external organization. With ABC Power providing core business expertise and SmartML providing machine learning expertise, respectively, an external Scrum master with Scrum and Agile framework expertise would be a valuable asset to the pilot project.

How development teams conduct meetings is dependent on the degree of colocation. With an external development team, most meetings will be virtual, including the 15-min daily Scrum meeting. In case of virtual meetings, artifacts such as product backlog, sprint backlog, and product increment can be adapted by using virtual collaboration tools.

The length of the sprint review, sprint retrospective, and sprint planning correlate with the length of the sprint. With back-to-back scheduling, an in-person meeting, where feasible, is ideal. The kind of communication, interaction, and feedback intended for these meetings are invaluable and will work significantly better in a traditional workshop environment. Typically, these meetings take place every 2 to 4 weeks, making in-person meetings achievable, even if team members are not colocated.

7.4.5 Stakeholders

Now we will shift gears from our ABC Power Ltd. example to a more general examination of Scrum applied to power companies. Outside the core team, you will have many stakeholders to collaborate with and the way you manage these stakeholders will be particularly important if your project competes with daily operations.

IT/OT and data management might be your most important collaborators. Machine learning projects are heavily reliant on data and your team needs access to this data. Enabling data availability with consideration of cybersecurity, compliance, access rights, data exchange protocols, and consumption will often be your most time-consuming task. Onboarding and integrating data security staff onto your team will save you future headaches and delays.

Further, getting management buy-in is necessary to execute these projects; however, it makes sense to differentiate between management levels. Generally, top management understands the added value of digitalization projects and provides support, whereas middle management often calls for more massaging. A large part of these machine learning projects comes down to change management, especially when deploying a Scrum approach (where management does not have a designated role). Thus it is imperative that middle management is aware, involved, and has ownership that has been communicated top down.

Further, it is important that your customers are fully aware of their role as customers. In a setting where a project is initiated from corporate headquarters into other operational organizations, the product owner might be someone who is not part of the actual operation's team. In fact, more often than not, this is the case.

As much as your project will be competing your customers' daily tasks and duties, their involvement will be required. Customer feedback will be critical to ensure user-friendliness and ultimately, user-acceptance. Your product could be the best from the perspective of the development team, but if your end customer is not using it, the product has failed. Your customer needs to be aware of his role, know how important his feedback is, and recognize that the development team and product owner are ultimately working to satisfy his needs.

7.5 Management of change and culture

New processes require change and change management, which can bring with it resistance. Consequently, your attention to the various stakeholders and your company's change culture will be crucial.

As previously mentioned, it is common for pilot projects to run in the form of rogue projects, i.e., in parallel to your core business activities. This approach has several advantages, including

- Setting context to test an assumption—The context of a pilot project gives you more liberties to experiment and test the underlying technology you want to implement to solve a certain problem. It is usually easier to get approval for smaller projects.
- Implementing quickly, managing risk, and failing fast—A quick test implementation will allow you to better manage risk and minimize cost. In case of failure, you will have failed fast and gracefully, rather than blowing up. Experimenting on a smaller scale educates you about what works and what does not without the level of risk of a larger project.
- Discovering dependencies—You will learn more about how your project interacts with various stakeholders as well as other devices, technologies, or data sources. Some of these stakeholders might not have been in your original scope. You will also learn more about the accuracy of your planned resource allocation.
- Fine-tuning the business case—The learnings from the pilot will allow you to better estimate your business case should you want to scale and implement your project on a larger scale after successfully completing the pilot.
- Getting feedback from the users, your customers—Next to understanding whether the technology itself works, your most important insights will be whether your idea will be accepted by the users.
- Refining your solution and spur new ideas—Based on the user feedback, you will be able to fine-tune your product, so it matches your customer's needs. This might result in new ideas and concepts. Further, smaller endeavors allow you to change direction quickly, if needed.
- Preparing for rollout—Experiential feedback and learning will be the best possible preparation for a later rollout and scaling of your product. A small-scale rollout is the perfect testing ground and makes a later, larger release less risky and stressful.

With pilots like these, you are actively engaging your colleagues as testers. For many, this kind of involvement will be a new experience, resulting in various levels of enthusiasm. Some will embrace this level of engagement and participation in the development of something novel, while others will not. In the context of Scrum, its requirements of inspection and adaptation, such an approach might be viewed as disorganized and chaotic. If you are used to Waterfall project management, shifting to Scrum-like methods may not be so intuitive. You will need to invest significant time into setting the stage by explaining the approach and coaching your colleagues on the importance of their roles.

Since you are not directly contributing to the core business, the pilot could also be perceived as a useless distraction. With potential headwinds like these, it is important to get the right mandate from the start. The more novel the approach, the more important an endorsement from higher management levels, such as CTO or CEO, will be. Management can be supportive by not only stressing the relevance to the organization for technical reasons, but also in the context of new ways of working and collaborating, including motivating the team to be brave and giving them the room and permission to fail. A basic understanding among top management of how Scrum and Agile works will give you yet another advantage.

Whatever the outcome of your pilot, you should not be afraid to share the results. With success or failure, there will be stories to tell. Be transparent, open, and honest. Use success stories that highlight your colleagues and customers and have them reflect, not only on the technical aspect, but on the way the team collaborated. Successful pilot projects can generate momentum that can be used to drive and multiply a solution into the organization.

7.6 Scaling—From pilot to product

With the completion of your pilot, hopefully you have a successful MVP. Assuming the product concept has been validated, you now have a better understanding of the technology and the risks and cost involved. Based on the positive feedback from your colleagues who serve as customers, you know you have a great product. With the product being operationally ready, it is now time to scale.

A great idea and productive pilot are only half of a successful product. Even more important is making sure you can scale your solution. Up until this point, the pilot has been a local, validation experiment—it is only after you scaled the solution to more assets that you can multiply and reap the full potential and benefits of the business case for the entire organization. Let us have a look at some important aspects of this process.

7.6.1 Take advantage of a platform

No matter which business units or assets you will scale your product to next, you will always need data and usually lots of it. Companywide, standardized data architecture will give your scaling efforts significant leverage. If your product only needs to learn to connect to one type of data source (e.g., a standardized data layer) your connectivity issues become close to plug and play, reducing scaling complexity significantly.

In the beginning of your scaling process you might find that the effort of getting relevant data connected to your data layer is a task greater than connecting your application. Once your product is up and running, it becomes a lot easier to replicate, whereas hooking up different data sources to your data layer can be tedious.

Considering data provision as part of your scaling requirements and as part of a global digitalization effort will give you additional leverage. Data availability will not only be relevant for your product, but hopefully, to future endeavors. These products will serve as business cases that will justify the data layer itself, since data provision alone does not create additional business value. Typically, data is moved from local data historians to global cloud solutions, facilitating access from anywhere in the world.

7.6.2 Establish a team and involve the assets

In the pilot phase, your team members might have been an ad hoc ensemble of colleagues that were sourced to execute the project. As you move from pilot to product, your engagement will consequently be more long term, including the creation of a self-sufficient, committed core team that can grow as needed.

Remember to involve the asset from the start. Involving the user at the beginning of each roll-out assures you meet requirements and that you understand expectations. Your products will fail if you do not create coownership.

7.6.3 Keep developing

Be prepared to keep developing your product. Your users will find bugs and request new features. The product will evolve with implementation and use, and unforeseen issues will need to be addressed. Staying open-minded with an eye toward improvement will help you meet these challenges.

As your product grows, you will need to ensure you scale its support. In the pilot phase, most issues will be handled by the development team. As the number of users increases, your capacity to support your solution will need to grow. Depending on the setup, this growth may be managed within your IT department or in conjunction with an external partner.

7.6.4 Involve UX expertise

While the pilot focuses on the technical feasibility of your product, you will have more time to fine-tune the user experience (UX) and design when you conceptualize the final product. In the end, your solution must not only work properly, but the user experience must be positive and sustainable. Again, the user's input will be critical to create an intuitive interface while maintaining high functionality. Finally, the design of the tool can also be aligned with the corporate design language of your company.

7.7 Further reading

This was but a glimpse into machine learning projects employing an Agile framework. If you want to dive deeper into any of the topics discussed earlier, we recommend the following reading list:

Schwaber, K., Sutherland, J. (2017): The Scrum Guide. https://www.scrumguides.org/.

Verheyen, G. (2013): Scrum—A Pocket Guide. Van Haren Publishing.

Schwaber, K. (2004): Agile Project Management with Scrum. Microsoft Press.

Cohn, M. (2004): User Stories Applied: For Agile Software Development. Addison-Wesley Professional.

McGreal, D., Jocham, R. (2018): The Professional Product Owner. Addison-Wesley Professional.

Lencioni, P. (2002): The Five Dysfunctions of a Team: A Leadership Fable. Jossey-Bass.

Adkins, L. (2010): Coaching Agile Teams: A Companion for Scrum Masters, Agile Coaches, and Project Managers in Transition. Addison-Wesley Professional.

Agilemanifesto, http://agilemanifesto.org/.

Fierro, D., Putino, S., Tirone, L., 2018. The Cynefin framework and the technical leadership: how to handle the complexity. CIISE 2017: INCOSE Italia Conference on Systems Engineering, vol. 2010.

Foxall, D., 2019. Scaling Your Product—Thriving in the Market. https://www.boldare.com/services/scaling-your-product-thriving-in-the-market/.

Jordan, J., 2018. Organizing Machine Learning Projects: Project Management Guidelines. https://www.jeremyjordan.me/ml-projects-guide/.

Kniberg, H., 2016. Making Sense of MVP (Minimum Viable Product)—And Why I Prefer Earliest Testable/Usable/Lovable.

https://blog.crisp.se/2016/01/25/henrikkniberg/making-sense-of-mvp.

Lynch, W., 2019. The Brief History of Scrum. https://medium.com/@warren2lynch/the-brief-of-history-of-scrum-15efb73b4701.

Newlands, M., 2017. The Perfect Product Is a Myth. Here's How to Scale the Almost-Perfect Product. https://www.entrepreneur.com/article/293010.

Singh, D., 2019. Things to Consider While Managing Machine Learning Projects. https://cloudxlab.com/blog/things-to-consider-while-managing-machine-learning-projects/.

Stacey, R.D., 2011. Strategic Management and Organizational Dynamics: The Challenge of Complexity, sixth ed. Pearson Education Limited, Harlow, Essex, England.

Wikipedia, https://en.wikipedia.org/wiki/Scrum_(software_development).

Zujus, A., 2019. AI Project Development—How Project Managers Should Prepare. https://www.toptal.com/project-managers/technical/ai-in-project-management.

Reference

Takeuchi, H., Nonaka, I., 1986. The New New Product Development Game. Harvard Business Review.

8

Machine learning-based PV power forecasting methods for electrical grid management and energy trading

Marco Pierro[a,b]**, David Moser**[b]**, and Cristina Cornaro**[a,c]

[a]*Department of Enterprise Engineering, University of Rome Tor Vergata, Rome, Italy*
[b]*EURAC Research, Bolzano, Italy.* [c]*CHOSE, University of Rome Tor Vergata, Rome, Italy*

Machine Learning and Data Science in the Power Generation Industry. https://doi.org/10.1016/B978-0-12-819742-4.00008-1
© 2021 Elsevier Inc. All rights reserved.

8.1 Introduction

A large share of Variable Renewable Energy generation (VRE) such as wind and solar is considered as one of the key strategies worldwide to decarbonize the energy sector (considering the trend toward electrification as well as for the heating, cooling, and transport sector) and thus reduce fossil fuel consumption.

On the other hand, VRE generation introduces new challenges for the stability of the electrical grid, both at local and national level, since it is perceived by Transmission and Distribution System Operators (TSOs and DSOs) as a reduction of the regional electric demand. Indeed, the majority of the wind[a] and solar produced electricity is locally produced and locally consumed, giving rise to a "load shadowing effect" (Rekinger et al., 2012), see Fig. 8.1. This residual electric demand becomes dependent on the VRE's inherent intermittency and stochastic variability (Perez et al., 2011; Perez and Hoff, 2013).

Due to the daily intermittency of solar irradiance (difference between daytime and nighttime hours), PV power generation can completely modify the daily shape of the electric load. During very overcast days the net load is almost equal to the total load with the typical increase until the night peak. During very clear sky days at noon (i.e., when the PV production is at its maximum) the net load is much lower than the total load, while a rapid increase in the evening (when the PV generation decreases in parallel to a growing electricity demand) appears. This results in a net load profile known as "duck curve" (Fig. 8.1). Therefore, during clear sky days, a remarkable increase of the net load power ramp

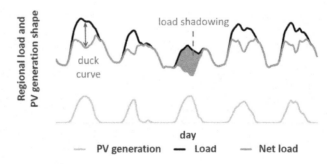

Fig. 8.1 Load, net load, and PV generation.

---

[a]Even if the majority of wind plants are not distributed since they are typically directly connected to HV transmission grid.

can occur. These higher power ramps imply a faster reduction of the energy supply between morning and noon followed by a faster increase of energy demand between noon and evening.

This change in load shape coupled with the solar-induced variability of the net load could potentially lead to many different issues (Kroposki et al., 2017; Stetz et al., 2015; Denholm et al., 2016) such as voltage fluctuation and voltage flicker, generation uncertainty, lack of stabilizing inertia needed to maintain the system frequency, unintentional islanding, increasing of power ramps, as well as transmission availability and reliability, reverse power flow, and energy imbalance problems. Reserve capacity is needed to control frequency changes that might arise from sudden, and sometimes unpredictable changes in generation or demand. Therefore the intermittent nature of the PV generation increases the amount of reserve that must be held for the dispatch process to accommodate the load-generation imbalance. To arrest the rate of change of frequency and stabilize the grid, the power reserve should be supplied by generators with a fast response level (such as hydroelectric or combined cycle power plants).

Furthermore, the instantaneous ratio between the PV generation and the electric demand could be much higher than the PV annual penetration (i.e., the fraction of annual electric demand provided by solar generation) especially during summer or during holidays when the demand is low. Thus, even with an apparently low PV penetration, the risk of negative impacts to the power grid (e.g., unintentional islanding and reverse power flow) cannot be excluded a priori.

Variable renewable energy penetration also affects the electricity markets. On one hand, it decreases the energy price on the day-ahead and intraday markets that is defined by the "merit order" rule, i.e., plants with lowest marginal costs are dispatched first. On the market sections, renewable sources in general, and wind and photovoltaics that have almost zero marginal costs of production in particular, are displacing traditional plants from the supply curve, reducing the energy price. On the other hand, VRE generation increases the energy dispatching prices on the energy balancing market, defined by the "pay as bid" rule. Indeed, the higher the imbalances between residual demand and supply, the higher the bids that the TSO is forced to accept. Furthermore, the increasing of reserve for grid stabilization has led to the definition of the capacity market, introducing additional capacity costs.

For these reasons, the concept of photovoltaic hosting capacity has been introduced as the maximum VRE capacity that can be integrated into the transmission/distribution grid with low

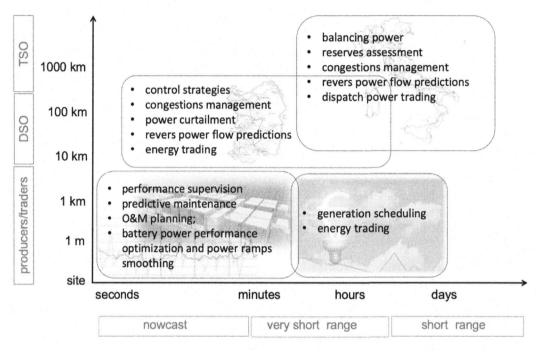

Fig. 8.2 Different use of PV power nowcast and forecast depending on the spatial and temporal scales.

impact on its stability, reliability, and security (Ding et al., 2016; Dubey et al., 2015; Jothibasu et al., 2016).

To mitigate the effects of high PV penetration and increase the hosting capacity, several strategies can be adopted by DSOs/TSO. In particular, solar generation nowcast and forecast is one of the first PV integration strategy that can be implemented because it does not imply any change in grid configuration or any installation of new electrical devices, and it is already available at a small fraction of grid operations' costs (Alet et al., 2016, 2017). Fig. 8.2 depicts the different use of PV power nowcast and forecast at different horizons and size of the controlled area.

In this chapter, we show how it is possible to predict the day-ahead photovoltaic generation using machine learning (ML) techniques both at the level of an individual photovoltaic system and at the national level for the Italian case study. We address some critical issues related to the implementation of ML forecasting models/methods. We show two possible applications of the day-ahead PV generation forecast for PV producers/traders and for the Italian TSO. The first deals with the day-ahead scheduling of the PV generation of utility-scale PV farms that PV producers/

traders should deliver to the Italian TSO. The second is related to the imbalance between supply and demand at national level, hence to the prediction of the Italian net load made by the Italian TSO to plan the reserves needed to resolve the day-ahead production uncertainty. For both cases, we provide the economic value of solar forecasting under the current imbalance regulatory framework.

The chapter is organized as follows:
1. We briefly describe the imbalance regulatory framework and balancing energy market in Italy.
2. We report the data used to carry out this study.
3. We describe the methodology and the implemented forecast model/method.
4. We show the economic value of single plant PV power forecast for producers/traders.
5. We show the economic value of the national PV generation forecasts for TSO/end-users.

8.2 Imbalance regulatory framework and balancing energy market in Italy

8.2.1 Transmission grid management: Market zones

In Italy there is only one TSO (Terna Spa) that manages the 72,900 km transmission grid. Terna is the largest transmission system operator in Europe and is in charge of the security and the stability of the national transmission grid. It prevents risks of power outages through constant monitoring of energy flows and a continuous service supply activity. To better ensure the security of the Italian electricity system, Terna divided Italy into different zones defined by physical energy transit limits of the national transition grid (Terna Spa (92/2019), 2019; Fig. 8.3).

These areas are aggregates of geographical and/or virtual zones, each characterized by an energy market and a zonal energy price. The areas are defined as follows:
- One national virtual zone affected by a limited generation capacity: Rossano.
- Nine international virtual zones: France, Switzerland, Austria, Slovenia, Corsica, Corsica AC, Montenegro, Malta, and Greece.
- Six geographical market zones covering the whole country, in which the majority of the energy needed to fulfill the national demand is exchanged: North (NORD), Center-North (CNOR), Center-South (CSUD), South (SUD), Sicily (SICI), and Sardinia (SARD).

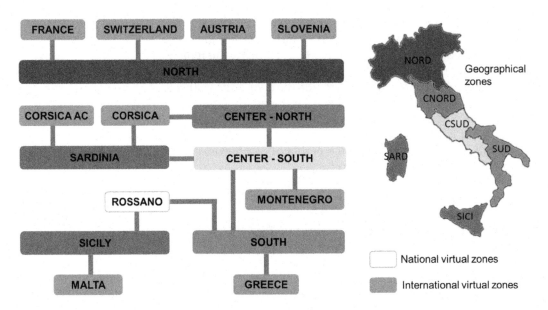

Fig. 8.3 Market zones. Modified from Terna Spa (92/2019), 2019. Individuazione della rete rilevante. Available from: https://www.terna.it/it/sistema-elettrico/mercato-elettrico/zome-mercato.

8.2.2 Imbalance regulation and economic valorization on BEM at national level

Terna manages the resolution of congestions and the mismatch between supply and demand to ensure the required system frequency and voltage. For this purpose, the TSO purchases/sells flexible power on the balancing energy market and issues generation/consumption dispatching orders to the Balancing Service Providers (BSP). The day before delivery, the TSO computes the reserves margin to compensate the imbalance and asks for energy bids on the dispatching market (MSD ex ante). Then, just before the day of delivery and during the day of delivery, the TSO asks for energy bids on the real-time balancing market (BM) to acquire the flexible capacity needed to overcome unpredictable events or errors in reserves estimation. Finally, on the day of delivery, the TSO accepts the bids on the MSD ex ante and MB and purchases/sells energy for upward/downward regulation to balance supply and demand.

The imbalance is defined as:

$$E_{\text{imb}}^{B}(Z, h) = E_{\text{generation}}^{\text{scheduled}}(Z, h) - E_{\text{netload}}^{\text{current}}(Z, h) \qquad (8.1)$$

where E_{imb}^{B} is the energy imbalance, $E_{\text{netload}}^{\text{current}}$ is the real energy demand, and $E_{\text{generation}}^{\text{scheduled}}$ is the generation scheduled on the day-ahead market (DA) and adjusted on intraday market (ID) (revised and corrected by Terna to avoid congestions). All these quantities are aggregated per zone (Z) and they refer to a specific hour of the day of delivery (h).

The annual imbalance volume is defined as the sum over the zones over time of the hourly energy imbalance:

$$\text{Volume} = \sum_Z \sum_{h=1}^N E_{\text{imb}}^B(Z, h) \qquad (8.2)$$

If the imbalance in a specific hour is negative (demand exceeds generation), Terna must implement an upward regulation ordering to the Production/Consumption Units (UP/UC allowed to provide dispatching services) to increase/decrease their generation/consumption. In this case, Terna purchases energy on the balancing markets (MSD and BM) at higher price than the DA market zonal price (P_Z^{DA}), accepting all the bids (needed to compensate the imbalance) on the basis of "pay as bid." The upward regulation price $(P_Z^{B\uparrow} \geq P_Z^{DA})$ is computed ex post as the zonal average of all the accepted bids of sale weighted with the corresponding exchanged energy (UPs sale price).

If the imbalance in a specific hour is positive (generation exceeds demand), Terna must implement a downward regulation ordering to the Production/Consumption Units to decrease/increase their generation/consumption. In this case, Terna sells (to UPs the not-produced energy) at lower price than the DA price. Also in this case, the downward regulation price $(P_Z^{B\downarrow} \leq P_Z^{DA})$ will result, ex post, from the zonal average of all the accepted bids of purchase weighted with the corresponding exchanged energy (UPs purchase price).

Fig. 8.4 reports the monthly average of the daily price on the DA market and the upward and downward prices on the MSD ex ante. The average DA price in 2015 in the North zone was 53 € per MWh while the yearly average of the two MSD ex ante prices were 104 and 23 € per MWh.

The imbalance costs are partially covered by the owners of utility-scale Production Unit (UP) that must pay the TSO in case of incorrect generation scheduling. The remaining major costs (mainly due to load, distributed VRE forecast errors, and unpredictable events) are borne by ratepayers, i.e., they are socialized.

It must be pointed out that the imbalance related to the accepted bids on MSD ex ante (which produce MSD imbalance costs) is defined as:

$$E_{\text{imb}}^{\text{MSD}}(Z, h) = E_{\text{generation}}^{\text{scheduled}}(Z, h) - E_{\text{netload}}^{\text{current}}(Z, h)$$
$$= E_{\text{netload}}^{\text{forecast}}(Z, h) - E_{\text{netload}}^{\text{current}}(Z, h) \qquad (8.3)$$

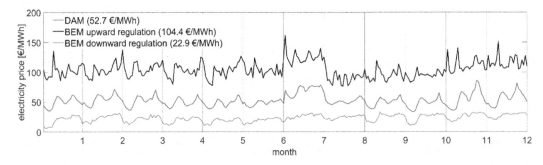

Fig. 8.4 Monthly average of the daily energy prices on the day-ahead energy market (DAM) and on the MSD ex ante (year 2015—North market zone). Modified from Terna Spa, Available from: https://www.terna.it.

$$E_{\text{netload}}^{\text{forecast}}(Z, h) = E_{\text{load}}^{\text{forecast}}(Z, h) - E_{\text{wind}}^{\text{forecast}}(Z, h) - E_{\text{PV}}^{\text{forecast}}(Z, h) \quad (8.4)$$

where for our specific calculation, $E_{\text{netload}}^{\text{forecast}}$ is the day-ahead net-load forecast resulting from the load, wind, and PV power predictions provided by the TSO forecast models ($E_{\text{load}}^{\text{forecast}}$, $E_{\text{wind}}^{\text{forecast}}$, $E_{\text{PV}}^{\text{forecast}}$).

Therefore the MSD imbalance volumes and costs result directly from the net load forecast errors and hence from the load, wind, and PV forecast model accuracies.

8.2.3 Production unit imbalance and economic valorization on balancing energy market

As mentioned earlier, only some UPs have to bear the costs of the imbalance between their real generation and the generation scheduled on the DA market.

According to the regulatory framework 281/2012/R/efr, 525/2014/R/EEL, 419/2017/R/eel of the Italian Authority for Electricity and Gas and the dispatch guidelines of Terna Spa (54/2018) (2018), the production units are divided into

- Relevant or not relevant.
- Allowed or not allowed to participate in the balancing markets providing dispatching services.

The relevant UP are single plants or aggregated plants in the same zone of the same typology (virtual production units) with a total capacity greater than 10 MW. For the quantification of the UP imbalance, the relevant units are managed at single plant level, while the nonrelevant units are managed in an aggregated manner at the zonal level.

Only the "Relevant" UPs (UPRs) have to deliver the day-ahead generation scheduling to the TSO and will incur in penalties for the imbalance with respect to their scheduled profile. In this case the generation imbalance is defined as:

$$E_{\text{imb}}^{\text{MSD}}(\text{UPR}, h) = E_{\text{generation}}^{\text{current}}(\text{UPR}, h) - E_{\text{generation}}^{\text{forecast}}(\text{UPR}, h) \qquad (8.5)$$

where $E_{\text{generation}}^{\text{current}}$ and $E_{\text{generation}}^{\text{forecast}}$ are the real and day-ahead scheduled generation of the UPR at the delivery hour (h). Therefore UPR negative imbalance means underproduction with respect to the scheduling while positive imbalance means overproduction.

Only a subset of the relevant production units is enabled to participate in the balancing markets depending on the flexibility and programmability of their power generation (Terna Spa (54/2018), 2018). In particular, variable renewable power plants such as wind and solar are currently not allowed to provide dispatching services.

To define the energy price of the "Relevant" UP imbalance, Terna divided Italy in two macro zones (Fig. 8.5). The first is the North market zone and the second is the remaining part of Italy.

The price of the generation imbalance of the "Relevant" UPs depends on the sign of the regional imbalance of the macro zone in which the UP is located. The macro zone's sign is computed by

Fig. 8.5 Macro zones. From Terna Spa (54/2018), 2018. Regole per il dispacciamento. Available from: https://download.terna.it/terna/0000/1066/54.PDF.

Table 8.1 Balancing energy market (BEM) "single pricing" imbalance valorization role for relevant production units not allowed to provide dispatching services (P_Z^{DA} is the DA zonal price; $P_Z^{B\downarrow}$ and $P_Z^{B\uparrow}$ are the BEM purchase/sale zonal prices).

		Sign of the imbalance of relevant production unit (market zone Z)	
		Overgeneration with respect to the day-ahead scheduling (+)	Undergeneration with respect to the day-ahead scheduling (−)
Sign of macro zonal imbalance	Over zonal generation (+)	UPR sells to TSO their imbalance at minimum BEM price $V_Z^\beta = \min\{P_Z^{DA}, P_Z^{B\downarrow}\}$ This price produces minimum imbalance cost	UPR purchases from TSO their imbalance at minimum BEM price $V_Z^\beta = \min\{P_Z^{DA}, P_Z^{B\downarrow}\}$ This price produces minimum imbalance revenue
	Under zonal generation (−)	UPR sells to TSO their imbalance at maximum BEM price $V_Z^\beta = \max\{P_Z^{DA}, P_Z^{B\uparrow}\}$ This price produces maximum imbalance revenue	UPR purchases from TSO their imbalance at maximum BEM price $V_Z^\beta = \max\{P_Z^{DA}, P_Z^{B\uparrow}\}$ This price produces maximum imbalance cost

Terna ex post (after the supply) using Eq. (8.1) so that the economic value of the UP imbalance is also known ex post.

The imbalance price of the "Relevant" UP that is not allowed to provide balancing services (such as wind and solar plants) is defined by the "single pricing" method (Table 8.1), i.e., the price on the balancing markets depends only on the sign of macro zonal imbalance. Due to the "single pricing" rule, the imbalance of relevant UPs not allowed to provide dispatching services (such as solar and wind farms) can be a cost or a revenue (Table 8.1).

The "single pricing" rule has been designed to take into account the effect of the imbalance of the single UPR on the overall imbalance of the macro zone (in which the UPR is located). If the signs of the imbalance are the same, the day-ahead scheduling errors of the UPR generation contribute to increase the imbalance of the macro zone and consequently it will be at a cost. On the contrary, if the imbalance signs are different, the scheduling errors contribute to reduce the macro zone imbalance so that the UPR imbalance has a positive impact resulting in an economic revenue.

In particular, the net imbalance energy value (NIV_Z) of relevant UPs, resulting from the DA incorrect scheduling, is the difference between the total energy income on the DA and MSD ex ante (EV_Z) and the ideal income that could be obtained on DA with no imbalance (perfect power forecast) (EV_Z(noimbalance)):

$$NIV_Z = EV_Z - EV_Z(\text{noimbalance})$$
$$= \sum_{h=1}^{N} E_{\text{imb}}^{\text{MSD}}(\text{UP}, h) * (V_Z^B(h) - P_Z^{\text{DA}}(h)) \qquad (8.6)$$

where $E_{\text{imb}}^{\text{MSD}}(\text{UP}, h)$ is the energy imbalance (Eq. 8.3), V_Z^B is the MSD imbalance price defined in Table 8.1, and P_Z^{DA} is the DA energy price.

The total energy income on the DA and MSD is the sum of the income of DA scheduling (EV_Z^{DA}) and imbalance value on MSD (EV_Z^{MSD}):

$$EV_Z = EV_Z^{\text{DA}} + EV_Z^{\text{MSD}}$$
$$= \sum_{h=1}^{N} \left[E_{\text{generation}}^{\text{forecast}}(\text{UP}, h) * P_Z^{\text{DA}} + E_{\text{imb}}^{\text{MSD}}(\text{UP}, h) * V_Z^B \right]) \quad (8.7)$$

while the ideal income that could be obtained on DA with no imbalance is

$$EV_Z(\text{noimbalance}) = \sum_{h=1}^{N} E_{\text{generation}}^{\text{real}}(\text{UP}, h) P_Z^{\text{DA}} \qquad (8.8)$$

Table 8.2 presents an example of a "single pricing" application computed using the average prices on the DA and the MSD markets reported in Fig. 8.4. The higher imbalance cost is achieved on the condition of UP undergeneration and negative macro zone imbalance. Nevertheless, remarkable incomes can be obtained when the sign of the UP imbalance is not the same as the sign of the zonal imbalance, i.e., when the UP over/under-generation helps to reduce the mismatch between supply and demand in the macro zone.

8.3 Data

8.3.1 National load and PV-wind generation data

To train and test the PV forecast model/method and its applications on single plant and national scales, we used DAM and MSD ex ante energy prices as well as Italian hourly load and generation measurements of the period 2014–16. The data were measured by Terna Spa and they are publicly available and downloadable from the Terna Spa (Website) (n.d.).

Table 8.2 Example of "single pricing" for the imbalance of relevant UPs not allowed to provide dispatching services (variable renewable generation plants).

Market hypothesis	UP generation hypothesis	UP generation hypothesis
$P_Z^{DA} = 52.7$ €/MWh $P_Z^{B\uparrow} = 104.4$ €/MWh $P_Z^{B\downarrow} = 22.9$ €/MWh	$E_{generation}^{forecast}(UP, h) = 70$ MWh $E_{generation}^{current}(UP, h) = 100$ MWh $E_{imb}^{MSD}(UP, h) = 30$ MWh	$E_{generation}^{forecast}(UP, h) = 100$ MWh $E_{generation}^{current}(UP, h) = 70$ MWh $E_{imb}^{MSD}(UP, h) = -30$ MWh

Single pricing	Overgeneration with respect to the day-ahead scheduling (+)	Undergeneration with respect to the day-ahead scheduling (−)
Over zonal generation (+)	$V_Z^B = \min\{P_Z^{DA}, P_Z^{B\downarrow}\} = 22.9$ €/MWh $EIV_Z = -894$ €	$V_Z^B = \min\{P_Z^{DA}, P_Z^{B\downarrow}\} = 22.9$ €/MWh $EIV_Z = 894$ €
Under zonal generation (−)	$V_Z^B = \min\{P_Z^{DA}, P_Z^{B\uparrow}\} = 104.4$ €/MWh $EIV_Z = 1551$ €	$V_Z^B = \min\{P_Z^{DA}, P_Z^{B\uparrow}\} = 104.4$ €/MWh $EIV_Z = -1551$ €

8.3.2 Satellite and numerical weather prediction data

The satellite and numerical weather prediction data have been supplied by the forecast provider Ideam Srl. Ideam is a start-up company that currently provides the meteorological predictions to major private Italian television networks and one of the most popular weather forecast websites.

A brief description of the provided data is as follows.

The satellite-derived irradiance (GHI) and ground temperature (T_{air}) data used for power estimation (0-h horizon now-cast) comes from the geostationary radiative fluxes products, under Météo-France responsibility. It was obtained by OSI SAF SSI algorithm applied to the satellite images provided by METEOSAT-9 (MSG-3) at 0° longitude, covering 60S-60N and 60W-60E, at 0.05° latitude-longitude. The data have a one-hour granularity and a spatial resolution of 12 km × 12 km covering the whole territory (1325 points). Each of the 1325 time series covers the period 2014–17.

The GHI and T_{air} prediction, used as forecast model inputs, were also generated by the Weather Research and Forecasting mesoscale model (WRF-ARW 3.8) with an initialization at 12 UTC, analyzing the 24-h forecasts starting from the following 00 UTC.

- Initial contour data for model initialization comes from GSF model output with a spatial resolution of 12 km × 12 km.

- Radiation scheme: "Rapid Radiative Transfer Model".
- Forecast horizon: 24-h.
- Temporal output resolution: 1-h.
- Spatial resolution: $12\,km \times 12\,km$ covering the whole country (1325 points).
- Each of the 1325 daily hindcasts time series covered the period 2014–17.
- The NWP irradiance data have been also postprocessed with an original Model Output Statistic to reach a "state of the art" accuracy.

8.4 ML techniques for PV power forecast

We developed two ML models to predict the PV generation of a single PV plant as well as to forecast the generation of the whole Italian PV fleet. Both forecast models have been trained on 1 year of historical data and tested on another year. Despite their complexity, once the ML algorithms have been trained, these prediction techniques are simple to use and require a very short computational time as they only involve algebraic operations between matrices and well-known equations.

It is also worth remarking that the global horizontal irradiance can be decomposed into two factors:

$$GHI = K_{CS}GHI_{CS} \qquad (8.9)$$

GHI_{CS} is the clear sky irradiance and represents the deterministic part of the solar irradiance that mainly depends on the sun's position. K_{CS} is the clear sky index and depicts the stochastic part of the solar irradiance related to the meteorological variability. Both the site and regional forecasting methods use the NWP predictions of the clear sky index as inputs. Indeed, dealing with ML techniques, a good practice is to remove deterministic information from the inputs. ML algorithms are needed to represent the relationship between the stochastic inputs and the expected output. Adding further information (described by known equations) as input will reduce their accuracy.

8.5 Economic value of the forecast of "relevant" PV plants generation

We developed a model to predict the day-ahead PV generation of "relevant" plants all over Italy, i.e., of optimally placed PV systems (30° tilted and south oriented) with capacity greater than 10 MWp. The observed generation of each single plant was further estimated by applying the same model used for the forecast but ingesting the current irradiance (retrieved from satellite data) instead of the irradiance forecast (from numerical weather prediction model). Then, we computed the annual volume and dispatching costs of the power imbalance between scheduled and observed generations of each "relevant" solar system all over the country. We also provide a concrete example of imbalance volumes and costs that can be obtained using our "state of the art" forecast to schedule the generation of the largest PV plant in Italy. In this way, we show to PV producers/traders, the PV forecast economic value on the balancing energy market.

8.5.1 Day-ahead forecasting model for single PV plant generation

The irradiance prediction obtained by WRF is affected by systematic errors. To partially correct the bias, we developed a neural network-based model (MOSNN) that ingests the WRF irradiance prediction and historical satellite-derived irradiance to provide a more accurate GHI forecast. Then, we developed a deterministic "physical based" model (DM) to map the irradiance prediction into the PV generation of an optimally tilted and oriented PV system. It consists of a chain of widely used semiempirical models: (1) a decomposition model (DISC) to retrieve the direct normal irradiance (DNI) and the diffuse irradiance (DHI) from the GHI data (Ineichen et al., 1992; Perez et al., 1990; Marion, 2015; Maxwell, 1987); (2) a transposition isotropic model (LJ) to retrieve the global irradiance on the plane of array of PV modules (GTI) (Liu and Jordan, 1961); and (3) the Sandia Array Performance Model (SAPM) to convert the incident irradiance GTI into PV generation (King et al., 2004). The forecasting model used to predict the generation of "Relevant" PV plants is summarized by the diagram in Fig. 8.6.

8.5.2 Results

The net imbalance value NIV_Z together with the imbalance values on the MSD market (EV_Z^{MSD}) and the total energy incomes on DA-MSD markets ($EV_Z = EV_Z^{DA} + EV_Z^{MSD}$) have been calculated, all over Italy, for the period 2014–16 (according to the hourly DA and MSD energy prices available on the Terna database). Therefore this work provides the annual incomes/costs that "relevant" PV plants can obtain on the energy markets, i.e., the economic value of the forecast of "relevant" PV plant generation.

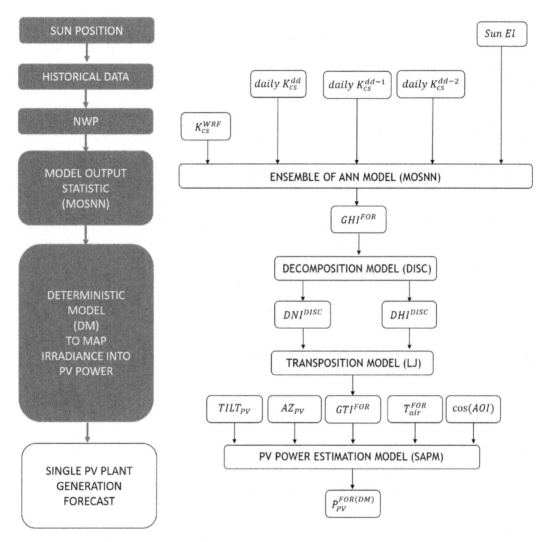

Fig. 8.6 Diagram of the model DM for PV single PY plant generation forecast. Input variables: WRF prediction of the clear sky index (K_{CS}^{WRF}), daily clear sky index of the first 3 days before the day of forecast (daily $K_{CS}^{(dd)}$, daily $K_{CS}^{(dd-1)}$, daily $K_{CS}^{(dd-2)}$), sun elevation (*SunEl*), tilt and orientation of the PV plant (*TILT*$_{PV}$, *AZ*$_{PV}$), air temperature forecast (T_{air}^{FOR}), and Angle of Incidence (AOI). Model output: PV plant generation forecast per unit of installed capacity ($P_{PV}^{FOR(DM)}$) expressed in MW/MWp.

Italy is divided into 20 regions (Fig. 8.7A). In average during the period 2014–16, there was a minimum yearly global horizontal solar radiation of 990 kWh/m^2 in the north alpine region and a maximum of 1900 kWh/m^2 in the south (Sicily) with a mean value of 1400 kWh/m^2 per year (Fig. 8.7B).

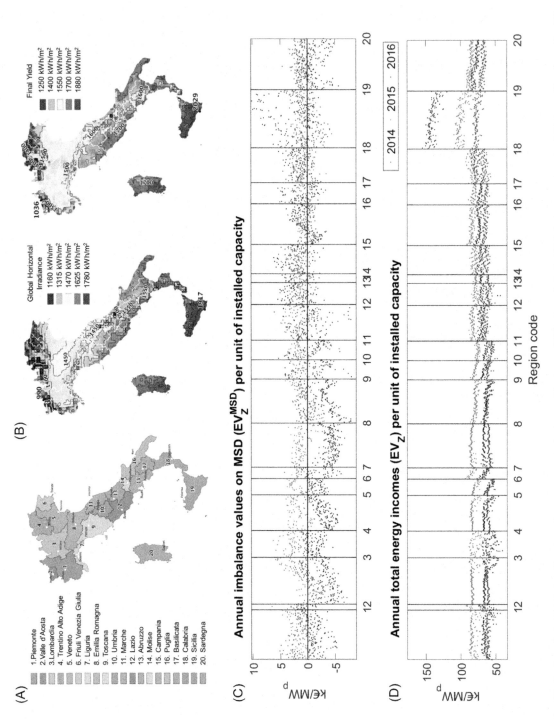

Fig. 8.7 Regions in Italy (A), irradiance and energy generation rate of "relevant" PV plants in Italy (B) imbalance values on MSD resulting from the PV generation forecast model (C); energy incomes on DA and MSD resulting from the PV generation forecast (D). Positive values represent incomes while negative values are costs.

The minimum generation rate (final yield) is around 1000 MWh/MWp in the north alpine region, while the maximum is around 2000 MWh/MWp in the south. The average generation rate of an optimal tilted and oriented PV system is around 1500 MWh/MWp (Fig. 8.7B).

Fig. 8.7C shows the imbalance value EV_Z^{MSD} (Eq. 8.7) on the MSD in different locations/years using our PV generation forecast. As reported in Table 8.1, the "single pricing" rule can produce, for "relevant" PV production units (UPR), positive (revenues) or negative (costs) imbalance values that, during 2014–16, range from 5 to −5 k€ per MWp of PV capacity per year.

There is a wide difference between the imbalance values obtained in different years. During 2014 and 2016, for most of the northern regions the imbalance is a cost (regions from 1 to 12) while in the southern regions it is mainly a profit (regions from 13 to 20). On the contrary, during 2015, energy imbalance generates a profit almost everywhere in Italy.

Fig. 8.7D shows the total energy incomes EV_Z on DA and MSD markets (Eq. 8.7) for "Relevant" PV producers in different locations/years.

These incomes range from 50 to 100 k€/MWp per year and the highest were obtained during the year 2015 with the only exception of Sicily. Indeed, in Sicily, during the year 2014 there was speculative action in the zonal energy market due to the grid bottleneck that did not allow the flow of additional cheaper power from the mainland.

The annual economic costs/revenues due to the errors in the PV generation scheduling are the net imbalance values NIV_Z (Eq. 8.6), computed as the difference between the total energy incomes on DA-MSD markets and the energy revenues that could be realized with a perfect forecast (no imbalance).

The net imbalance values are losses in all Italian regions with the exception of Lazio, Campania, Calabria, and Sardinia where there are some locations in which the imbalance could even yield profits (Fig. 8.8A).

The maximum economic losses are in the northern zone with a zonal average of −2.8 k€/MWp and a range of −4.9 to −0.7 k€/MWp while the minimum losses are in Sardinia with zonal average of −0.6 k€/MWp and a range of −1.7 to 0.5 k€/MWp.

Therefore, in Italy the value of the PV power forecast of "relevant" systems with the current "single pricing" rule ranges from a revenue of 0.5 k€/MWp to an imbalance cost of 4.9 k€/MWp, depending on the plant's location.

To better understand these numbers, we consider, for example, the largest solar farm in Italy (and in Europe). This is a 84 MWp photovoltaic power station at Montalto di Castro in Lazio Region

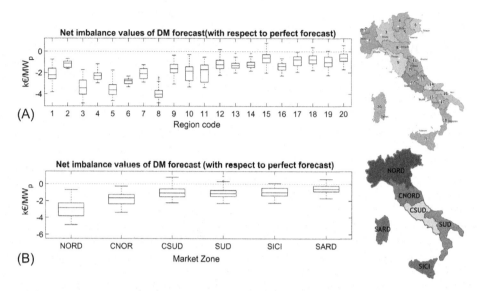

Fig. 8.8 Net imbalance values (averaged on 2014–16) aggregated by regions and market zones.

(region 12). The project was developed by the independent developer SunRay that was later acquired by SunPower[b] (Fig. 8.9).

The 2014–16 average generation of this power station (computed by our Satellite data) was 145 GWh/year while the energy imbalance around 41 GWh/year, around 28% of the generation (Table 8.3). According to Fig. 8.7, the imbalance value and the total incomes are very variable year by year, depending on the energy market prices.

The average imbalance value was −219 k€/year, corresponding to a mean energy MSD price of −5.5 €/MWh of imbalanced energy (negative values are costs for PV producers/traders). The average energy income on MSD and DA markets is 8091 k€/year (37 times the imbalance costs) with an average price of 57.4 €/MWh of produced energy (comparable with the National Unique Price, Fig. 8.4).

Finally, the net imbalance cost, i.e., the value of the PV forecast, was around −160 k€/year, corresponding to an imbalance price of −1.1 €/MWh of produced energy. Even if the imbalance costs are significantly smaller than the energy income, these costs for "relevant" solar farms are not negligible and are much higher than the price of an accurate PV power prediction delivered by a forecast provider.

[b] https://en.wikipedia.org/wiki/Montalto_di_Castro_Photovoltaic_Power_Station.

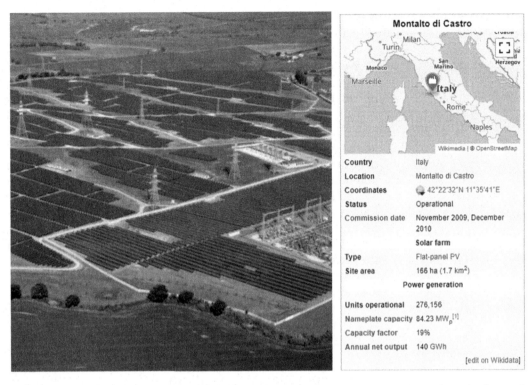

Fig. 8.9 Largest PV farm in Italy. Image from SMA Solar, https://www.sma-italia.com/prodotti/referenze/montalto-di-castro.html; data from Wikipedia: https://en.wikipedia.org/wiki/Montalto_di_Castro_Photovoltaic_Power_Station

Table 8.3 Generation, energy imbalance, and economic values on the energy markets.

Year	Final yield (MWh/MWp)	Energy generation (GWh)	Absolute energy imbalance (GWh)	Imbalance energy value on MSD (EV_Z^{MSD}) (k€)	Total energy income (EV_Z) (k€)	Net imbalance value (NIV_Z) (k€)
2014	1651	139	43	−485	10,131	−229
2015	1742	147	41	−179	7993	−124
2016	1762	148	40	6	6150	−122
Avg.	1718	145	41	−219	8091	−158

8.6 Economic value of PV forecast at national level

We further built up a method to forecast the Italian PV generation directly and we used this forecast to predict the national residual load of the next day (net load):

$$P_{\text{Netload}}^{\text{FOR}} = P_{\text{Load}}^{\text{FOR}} - P_{\text{Wind}}^{\text{FOR}} - P_n P_{\text{PV}}^{\text{FOR}} \tag{8.10}$$

where $P_{\text{Load}}^{\text{FOR}}$ is the day-ahead prediction of the national electric demand; $P_{\text{Wind}}^{\text{FOR}}$, $P_n P_{\text{PV}}^{\text{FOR}}$ are the day-ahead forecast of the wind and photovoltaic national generation; P_n is the PV installed capacity and $P_{\text{Netload}}^{\text{FOR}}$ is the net load prediction, i.e., the scheduling of the Italian generation that must be supplied by other programmable RE and fossil plants. The imbalance volume and costs are then computed using the observed residual load and compared with the imbalance volume and costs obtained by the Italian TSO forecast method. Thus we supply a real case study of PV power forecast application in support of TSO/DSO grid management. The description of the load and wind forecast models is out of the scope of this work.

8.6.1 Forecasting upscaling method for regional PV generation

The starting point for regional PV power forecasts is the so-called bottom-up strategy. It consists in the estimation or forecast of all the distributed PV plants in the considered area. Nevertheless, ongoing research is focused on upscaling methods that allow the estimation and forecast of distributed power of aggregates of PV plants through simplified approaches that reduce the computational effort and require less information about the PV fleet (Fonseca et al., 2015). Since the performance of each site forecast only slightly affects the performance of regional prediction, upscaling methods can achieve similar accuracy as bottom-up approaches.

Different upscaling strategies have been tested (Lorenz et al., 2008, 2011, 2012; Fonseca et al., 2014; Wolff et al., 2016). Here we used a "virtual power plant" approach, i.e., we considered the PV generation in the controlled area as it was produced by a virtual PV plant. Then, the power output of this virtual plant is directly forecasted by machine learning algorithms as in Zamo et al. (2014).

Upscaling methods usually consist of different steps: input preprocessing, forecast model itself, and eventually output postprocessing.

The preprocessing procedure aims to spatially aggregate and/or select the NWP data to reduce the dimensionality of the input features of the forecasting model as well as to retrieve other relevant information from the historical data. In particular, the reduction of the input dimensionality is essential when machine learning techniques (ML) are used to provide the power forecast. Indeed, for any ML model (neural networks, support vector regression, quantile regression forests, etc.) the ingestions of highly correlated features can completely hinder the performance since many correlated features do not bring any additional information while the model complexity increases dramatically with the number of inputs. The forecast model is the core of the method; it should map the NWP products together with other relevant information into regional PV power. Finally, the postprocessing provides a further refinement of the forecast model output, correcting systematic errors due to missing/incorrect input information.

We developed an upscaling method based on a ML forecast approach to directly predict the PV generation for the whole of Italy. The procedure, depicted in Fig. 8.10, can be described as follows:

- Preprocessing

    Preprocessing has two different aims. The first is to retrieve the optimal "equivalent" plane of the array of PV modules (POA) of the regional "virtual" PV system. We use the previously described DM model to transform the satellite-derived GHI (averaged all over Italy) into the power generated by a virtual PV system with a specific POA. Then, we optimize the POA minimizing the error between the DM output and the measured Italian PV generation (Pierro et al., 2017). The optimal "equivalent" tilted and orientation angles are retrieved using the training year data and then used to predict the equivalent Angle of Incidence during the test year (AOI^{FOR}).

    The second aim is to spatially aggregate the irradiance and temperature NWP computed in 1325 locations all over the country, reducing the input feature dimensionality. The NWP global horizontal irradiance was previously corrected by the MOSNN. Different aggregation methods were tested, all yielding similar forecast accuracy. Nevertheless, the outperforming forecast is the result of two different kinds of aggregation: the average of the 1325 predicted irradiances and temperatures all over Italy (GHI^{FOR} and T_{air}^{FOR}) and the average and standard

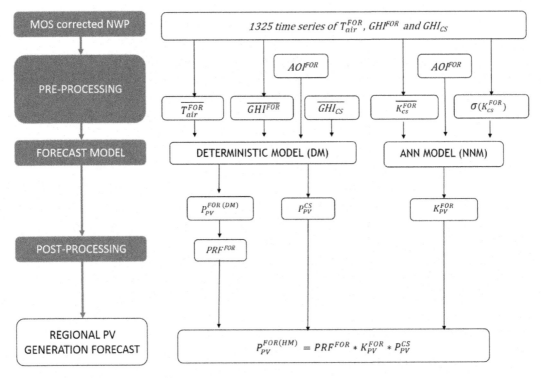

Fig. 8.10 Diagram of upscaling method (HM) for national PV generation forecast. Input variables: equivalent angle of incidence AOI^{FOR}, Italian mean values of the NWP model forecasts of the irradiance, temperature and clear sky irradiance ($\overline{GHI^{FOR}}$, $\overline{T_{air}^{FOR}}$ $\overline{GHI_{CS}}$), average and standard deviation on the Italian market zones of the day ahead forecast of the clear sky index ($\overline{K_{CS}^{FOR}}$, $\sigma(K_{CS}^{FOR})$). Model output: Italian PV generation forecast per unit of installed capacity (P_{PV}^{FOR} $^{(HM)}$) expressed in MW/MWp.

deviation of the 1325 predicted clear sky indices over the six market zones ($\overline{K_{CS}^{FOR}}$ and $\sigma(K_{CS}^{FOR})$). In this way, we reduced the dimensionality of the input features from 2650 (1325 irradiance and temperature NWP data) to 14.

- Forecast model

The day-ahead forecast of the Italian PV generation is delivered by a hybrid model (HM) based on the parallel working of a deterministic model (DM), and an ensemble of Neural Network models (NNM). DM (described in the previous section) is used to provide a first prediction of the average Italian PV generation ($P^{FOR(DM)}$) and to compute the clear sky PV generation, i.e., the PV power that could have been generated in a clear sky condition (P_{PV}^{CS}). NNM is employed to forecast the PV clear sky index (K_{PV}^{FOR}) that depicts the changes in PV generation with respect to the generation expected during clear sky conditions.

- Postprocessing

The postprocessing procedure aims to predict a Performance Correction Factor (PRF^{FOR}) that rescales the deterministic model (DM) output. This daily rescaling factor accounts for errors in capacity estimation, degradation, and AC losses not included in the physical model DM and for NWP bias errors. The performance correction factor is computed using the DM forecast ($P_{PV}^{FOR(DM)}$) and the PV observed generation (P_{PV}^{OBS}) of the current day (dd):

$$PRF^{FOR}(dd+1) = \sum\nolimits_{h=1,24} P_{PV}^{OBS}(h|\, dd) / \sum\nolimits_{h+1,24} P_{PV}^{FOR(DM)}(h|\, dd)$$

(8.11)

The Italian PV generation normalized by the PV installed capacity (MW/MWp) provided by the upscaling method (HM) is then obtained as follows:

$$P_{PV}^{FOR(HM)} = PRF^{FOR}\, K_{PV}^{FOR} * P_{PV}^{CS}$$ (8.12)

This method takes into account both the spatial variability of the irradiance (since it inputs the mean and the standard deviation of the clear sky indices in each market zone) and the nonuniform distribution of PV capacity (since the model will automatically weigh market zones inputs according to PV capacity of each zone). The model is trained on year 2015 and tested on year 2016.

8.6.2 Results

As mentioned earlier, our upscaling approach considers the Italian generation as if it was produced by a virtual power plant.

The first step of our preprocessing procedure consists in retrieving an "equivalent" plane of array (POA) of this virtual power plant.

The "equivalent" POA is retrieved by the deterministic minimization of a nontrivial cost function and corresponds to the tilt and the orientation angles for which the deterministic model (DM) (that ingests the national average values of the satellite derived GHI and T_{air}) provides the outperforming estimation of the Italian PV generation.

Fig. 8.11A shows that the cost-function is a convex function and reports the convergence of the minimization algorithm to the "equivalent" POA of the year 2016. Fig. 8.11B reports the agreement between the PV power estimation and the observed national PV generation.

Table 8.4 reports the "equivalent" angles retrieved in different years showing that all are south oriented and around 10 degrees tilted. This nonoptimal tilt (optimal tilt is 30 degrees) could

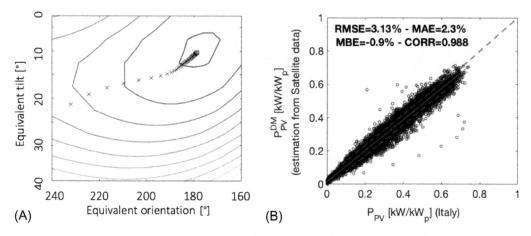

Fig. 8.11 Convergence of the minimization algorithm to retrieve the "equivalent" POA—the isolines show the POA that leads to the same cost-function value—(A); National power generation obtained by the deterministic model DM using the "equivalent" POA and the Italian average of the satellite-derived GHI and T_{air} (B). Figure (B) also reports the main key performance indexes used to evaluate the accuracy: Pearson correlation (CORR), root mean square error (RMSE), mean absolute error (MAE), and mean bias error (MBE). The first KPI is dimensionless while the others are in % of the PV installed capacity (P_n).

Table 8.4 "equivalent" POA retrieved in different years.

Year	2014	2015	2016
"equivalent" orientation (degrees)	184.7	182.8	181.1
"equivalent" tilt (degrees)	7.9	8.0	10.0

depend on the large capacity of large PV systems installed on industrial buildings[c] with almost horizontal roofs and/or multi-tilted PV plants.

Furthermore, as expected, the "equivalent" POA does not notably change year by year since new installations can only slightly modify these average "equivalent" angles. Therefore we used the "equivalent" POA retrieved using satellite and observation data of 2015 for the prediction of the PV generation of year 2016.

Fig. 8.11B shows that the estimation mean absolute error (MAE) is around 2% of the installed capacity (P_n) proving that

[c] In Italy the 25% of the installed power is concentrated in the industrial sector, in particular in PV systems with capacity between 200 kW and 1 MW.

the deterministic model is quite accurate, and it could be used to predict the National PV generation. Nevertheless, the deterministic approach does not consider the irradiance and temperature spatial variability and the spatial nonuniformity of the PV capacity all over the country. Indeed, the DM model uses only the national averages of GHI and T_{air}.

To account for these factors, we developed the neural network model (NNM) that ingests the NWP of clear sky index (K_{CS}^{FOR}— Eq. 8.9) all over the country and automatically weigh these values to better fit the power output. Thus this weight should also be related to the PV capacity spatial distribution. However, the dimensionality of the NN input features (K_{CS}^{FOR} of 1325 locations) should be reduced to provide an outperforming forecast.

We tried three different input aggregations methods:

1. Clustering the K_{CS} on the basis of location and irradiance values. In particular, we decided to use 20 clusters, i.e., 20 GHI time series located on the cluster centroids (Fig. 8.12A). Indeed, more than 20 clusters do not notably reduce the dispersion function. The NN model ingests 20 different K_{CS}^{FOR} values.

2. Applying a Principal Component Analysis to the K_{CS}^{FOR} retaining only the components that correspond to the higher eigenvalues. Fig. 8.12B shows the first 20 values of the 1325 K_{CS}^{FOR} eigenvalues (in a decreasing order). It is worth noting that the first eigenvalue accounts for the majority of the variance while the first 6 eigenvalues represent 90% of the variance. We chose to use only six principal components as inputs to the NN model since more input components resulted in lower forecast accuracy. Fig. 8.12(b1) reports an example of 1325 K_{CS} trends of 5 days (sun hours only). Fig. 8.12(b2) reports the 1325 trends of the K_{CS} corresponding only to the first PCA component, showing that this component is just depicting the daily shape of the K_{CS} all over Italy.

3. We just used the six market zones' average of the 1325 K_{CS}^{FOR} values together with the corresponding standard deviation. Indeed, the average depicts the mean weather variability between market zones while the standard deviation accounts for the weather variability inside each market zone. Thus the NN model uses 12 K_{CS}^{FOR} derived inputs instead of 1325.

Table 8.5 reports the accuracy of our hybrid method using the three preprocessing procedures to reduce the NN input features. All the three procedures lead to a NNM PV power forecast with very similar accuracy. Nevertheless, the method that makes use of average and standard deviation of the 1325 K_{CS}^{FOR} over each

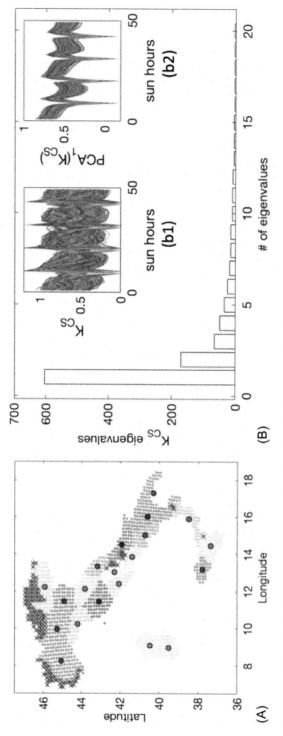

Fig. 8.12 Clustering of the 1325 Italian GHI time series based on the locations and irradiance values.

Table 8.5 Key performance indexes (KPIs) of the NNM forecast with the different preprocessing method.

	CORR	RMSE (% of Pn)	MAE (% of Pn)	MBE (% of Pn)
NNM (6 market zones)	0.98	3.67	2.45	0.3
NNM1 (20 clusters)	0.98	3.80	2.51	0.32
NNM2 (6 PCA)	0.97	3.86	2.57	0.35

The KPIs are Pearson correlation (CORR), root mean square error (RMSE), mean absolute error (MAE), and mean bias error (MBE). The first KPI is dimensionless while the others are in % of the PV installed capacity (P_n).

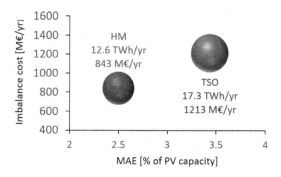

Fig. 8.13 Solar prediction accuracy (MAE), imbalance costs, and volume achieved during 2016 by the current TSO forecast (Terna Spa, 2017) and by our HM forecast.

market zone is the outperforming one. From now on, we will refer to this outperforming model as NNM.

Finally, using this outperforming preprocessing procedure and the NMM model and the postprocessing described in the previous section, we set up our PV power forecast upscaling method (HM) and subsequently the net load forecast (Eq. 8.10).

Fig. 8.13 compares the solar prediction accuracy (MAE), imbalance cost for the Italian TSO/ratepayers, and the imbalance volume achieved during 2016 by the current TSO forecast (Terna Spa, 2017) and by our ML based forecast (HM).

Improving the PV forecast accuracy reduces the imbalance volume and cost by 27.4% and 30.5%, respectively, with respect to the 17.3 TWh/year and 1213 M€/year obtained by the TSO forecast during the year 2016 (Terna Spa, 2017). The imbalance volume and cost obtained by our forecast are 12.6 TWh/year and 843 M€/year. Our HM solar forecast obtains MAE of 2.5% while the current TSO model achieves a MAE of 3.4%. This means that the value for Italian ratepayers of 1% MAE reduction is around 400 million of Euros.

8.7 Conclusions

We showed how ML techniques can be used for solar power forecast. We highlight two critical issues that need attention when implementing data-driven methods. The first consists in removing the deterministic and predictable information from the ML input features. This will improve the effectiveness of the ML algorithms in reconstructing the unknown relationship between the residual stochastic inputs and the expected output. The second is the use of preprocessing procedures to reduce input features dimensionality, isolating the essential (uncorrelated) information. This will reduce the complexity of the model, enhancing forecast accuracy.

We report two possible applications of solar forecasting to reduce solar-induced imbalance and improve the system stability for the case of high solar penetration. Then, we compute the economic value of solar forecasts on the dispatching energy markets for both applications.

For producers and traders that manage "relevant" PV farms, the economic value of accurate forecasts ranges from a revenues of 500 €/year per MWp of installed capacity to an imbalance loss of 5000 €/year per MWp, depending on the plant location. For the Italian TSO the value of 1% of solar forecast accuracy improvement is 400 million Euros, i.e., around 20 €/year per ratepayer.[d]

References

Alet, P.-J., Efthymiou, V., Graditi, G., Henze, N., Juel, M., Moser, D., Nemac, F., Pierro, M., Rikos, E., Tselepis, S., Yang, G., 2016. Forecasting and observability: critical technologies for system operations with high PV penetration. In: Proceedings of the 32nd European Photovoltaic Solar Energy Conference and Exhibition, Paris, June.

Alet, P.-J., Efthymiou, V., Graditi, G., Henze, N., Juel, M., Moser, D., Nemac, F., Pierro, M., Rikos, E., Tselepis, S., Yang, G., 2017. Assessing the Need for Better Forecasting and Obsevability of PV. European Technology and Innovation Platform.

Denholm, P., Clark, K., O'Connell, M., 2016. Emerging Issues and Challenges With Integrating High Levels of Solar Into the Electrical Generation and Transmission Systems. National Renewable Energy Laboratory, Golden, CO., NREL/TP-6A20-65800http://www.nrel.gov/docs/fy16osti/65800.pdf.

---

[d]There are about 20 million ratepayers in Italy.

Ding, F., Mather, B., Gotseff, P., 2016. Technologies to Increase PV Hosting Capacity in Distribution Feeders Preprint Technologies to Increase PV Hosting Capacity in Distribution Feeders. IEEE Power and Energy Society General Meeting (PESGM), Boston, MA, pp. 1–5.

Dubey, D., Santoso, S., Maitra, A., 2015. Understanding photovoltaic hosting capacity of distribution circuits. In: IEEE Power & Energy Society General Meeting, Denver, CO, USA.

Fonseca, J.G., Oozeki, T., Ohtake, H., Shimose, K.-I., Takashima, T., Ogimoto, K., 2014. Regional forecasts and smoothing effect of photovoltaic power generation in Japan: an approach with principal component analysis. Renew. Energy 68, 403–413.

Fonseca, J., Oozeki, T., Ohtake, H., Takashima, T., Ogimoto, K., 2015. Regional forecasts of photovoltaic power generation according to different data availability scenarios: a study of four methods. Prog. Photovolt. Res. Appl. 23 (10), 1203–1218.

Ineichen, P., Perez, R., Seal, R., Zalenka, A., 1992. Dynamic global-to-direct irradiance conversion models. ASHRAE Trans. 98, 354–369.

Jothibasu, S., Dubey, A., Santoso, S., 2016. Integrating Photovoltaic Generation. Enery Institute of the University of Texas, Austin.

King, D., Kratochvil, J., Boyson, W., 2004. Photovoltaic Array Performance Model. Sandia National Laboratories, Albuquerque, NM, USA.

Kroposki, B., Johnson, B., Zhang, Y., Gevorgian, V., Denholm, P., Hodge, B., Hannegan, B., 2017. Achieving a 100% renewable grid: operating electric power systems with extremely high levels of variable renewable energy. IEEE Power Energ. Mag. 15 (2), 61–73.

Liu, B., Jordan, R.C., 1961. Daily insolation on surfaces tilted towards equator. ASHRAE Trans. 10, 526–541.

Lorenz, E., Hurka, J., Karampela, G., Heinemann, D., Beyer, H.S., 2008. Qualified forecast of enesemble power production by spatially dispersed gri-connected PV systems. In: 23rd EU PVSEC section 5AO.8.6.

Lorenz, E., Scheidsteger, T., Hurka, J., 2011. Regional PV power prediction for improved grid integration. Prog. Photovolt. Res. Appl. 19, 757–771.

Lorenz, E., Heinemann, D., Kurz, C., 2012. Local and regional photovoltaic power prediction for large scale grid integration: assessment of a new algorithm for snow detection. Prog. Photovolt. 20 (6), 760–769.

Marion, B., 2015. A model for deriving the direct normal and diffuse horizontal irradiance from the global tilted irradiance. Sol. Energy 122, 1037–1046.

Maxwell, E., 1987. A Quasi-Physical Model for Converting Hourly Global Horizontal to Direct Normal Insolation. Solar Energy Research Institute (SERI), Golden, CO, SERI/TR-215-3087 DE87012273.

Perez, R., Hoff, T., 2013. Solar resource variability. In: Solar Energy Forecasting and Resource Assessment. first ed. Academic Press, Waltham, pp. 133–148.

Perez, R., Seals, R., Zelenka, A., Ineichen, P., 1990. Climatic evaluation of models that predict hourly direct irradiance from hourly global irradiance: prospects for performance improvements. Sol. Energy 44, 99–108.

Perez, R., Hoff, T., Kivalow, S., 2011. Spatial and temporal characteristics of solar radiation variability. In: International Solar Energy (ISES) World Congress, Kassel, Germany.

Pierro, M., Belluardo, G., Ingenhoven, P., Cornaro, C., Moser, D., 2017. Inferring the performance ratio of PV systems distributed in an region: a real-case study in South Tyrol. In: 44th IEEE Photovoltaic Specialists Conference (PVSC), Washington, DC, USA.

Rekinger, M., Theologitis, I.-T., Masson, G., Latour, M., Biancardi, D., Roesch, A., Concas, G., Basso, P., 2012. Connecting the Sun. EPIA, Belgium.

Stetz, T., Von Appen, J., Niedermeyer, F., Scheibner, G., Sikora, R., Braun, M., 2015. Twilight of the grids: the impact of distributed solar on Germany's energy transition. IEEE Power Energ. Mag. 13 (2), 50–61.

Terna Spa, 2017. https://www.terna.it/it-it/sistemaelettrico/dispacciamento/datiesercizio/rapportomensile.aspx/Rapporto_mensile sul sistema elettico_gennaio2017.

Terna Spa (54/2018), 2018. Regole per il dispacciamento. Available from:https://download.terna.it/terna/0000/1066/54.PDF.

Terna Spa (92/2019), 2019. Individuazione della rete rilevante. Available from: https://www.terna.it/it/sistema-elettrico/mercato-elettrico/zome-mercato.

Terna Spa (Website). Available from:https://www.terna.it.

Wolff, B., Kühnert, J., Lorenz, E., Kramer, O., Heinemann, D., 2016. Comparing support vector regression for PV power forecasting to a physical modeling approach using measurement, numerical weather prediction, and cloud motion data. Sol. Energy 135, 197–208.

Zamo, M., Mestre, O., Arbogast, P., Pannekoucke, O., 2014. A benchmark of statistical regression methods for short-term forecasting of photovoltaic electricity production part I: deterministic forecast of hourly production. Sol. Energy 105, 792–803.

9

Electrical consumption forecasting in hospital facilities

A. Bagnasco[a,b], F. Fresi[c], M. Saviozzi[a], F. Silvestro[a], and A. Vinci[b]

[a]*Department of Electrical, Electronic, Telecommunication Engineering and Naval Architecture (DITEN), University of Genova, Genova, Italy*
[b]*IESolutions Soluzioni Intelligenti per l'Energia, Genova, Italy*
[c]*Gruppo Humanitas, Clinica Cellini, Torino, Italy*

The topic of **energy efficiency** applied to **buildings** represents one of the key aspects in today's international energy policies. **Emissions reduction** and the achievement of the targets set by the Kyoto Protocol are becoming a fundamental concern in the work of engineers and technicians operating in the energy management field. Optimal energy management practices need to deal with uncertainties in generation and demand. Hence the development of reliable forecasting methods is an important priority and area of research in electric energy systems. This chapter presents a load forecasting model and the way it was applied to a real case study in forecasting the electrical consumption of the Cellini medical clinic of Turin, Italy. The model can be easily integrated into a Building Management System or into a real-time monitoring system. The load forecasting is performed through

Machine Learning and Data Science in the Power Generation Industry. https://doi.org/10.1016/B978-0-12-819742-4.00009-3
© 2021 Elsevier Inc. All rights reserved.

the implementation of an artificial neural network (ANN). The proposed multilayer perceptron ANN, based on a backpropagation training algorithm, is able to take as inputs: loads, type of day (e.g., weekday/holiday), time of the day, and weather data. This work focuses on providing a detailed analysis and an innovative formal procedure for the selection of all ANN parameters.

9.1 Introduction

Large buildings certainly represent some of the great consumers of electric energy. It is estimated that the worldwide energy demand of buildings is about 32% of total energy consumption. In Europe, buildings are responsible for 36% of CO_2 emissions (OECD/IEA, International Energy Agency, 2014). The need to achieve Europe's 2020 energy targets, i.e., 20% improvement in the EU's energy efficiency and 20% reduction in EU greenhouse gas emissions from 1990 levels (RTE: rseau de transport d'lectricit, Le bilan lectrique francais 2010, 2011), together with the need to ensure operational goals with the minimum energy cost and environmental impact, has contributed to the interest growth in Building Management Systems (BMS). These systems can contribute to achieving energy savings and thus cost savings (Doukas et al., 2007; Missaoui et al., 2014).

Load forecasting can be a useful function of a generic BMS, as well as representing common problems and key factors in electrical distribution systems. Several studies have been presented on short-term load forecasting (Abdel-Aal, 2004; Ying et al., 2010; Amjady et al., 2010), often using statistical models (Yan et al., 2012; Hong et al., 2014), regression methods (Ceperic et al., 2013), state-space methods (Lu and Shi, 2012), evolutionary programming (Amjady and Keyna, 2009), fuzzy systems (Khosravi and Nahavandi, 2014), and artificial neural networks (ANN) (Jetcheva et al., 2014; Webberley and Gao, 2013; Krstic et al., 2014). Among these algorithms, ANN has received more attention because of its clear model, ease of implementation, and good performance.

Load forecasting can provide vital information for energy evaluation within a single utility and for efficiency purposes, particularly when medium and low voltage energy distribution systems are considered. Load forecasting is also advantageous for system economics. In fact, it can provide valuable information to detect opportunities and risks in advance. The new energy market and the smart grid paradigm ask for both better load management policies and more reliable forecasts from single end-users up to system scale. The necessity of coordinating uncertain renewable generation with actual demand is becoming more and more important (Grillo et al., 2012).

Load forecasting with lead times from a few minutes to several days is an essential tool to effectively manage buildings and facilities. The correct forecast of the interchange between the site and the main grid is a key issue for the new paradigm of Smart Grid Management since a correct forecast will lead to improved control of the network. This may be of great importance in microgrid configurations, which are expected to play a larger role in future power systems.

Moreover, if the work environment is a hospital facility, where a continuous energy and electricity flow is required, all the aspects previously listed are amplified and of greater importance. A **hospital** can be defined as a highly complex organization under a functional, technological, economic, managerial, and procedural standpoint. A modern hospital can be compared to an industrial plant for the multiplicity and the type of its functions and tasks. Electric energy is the essential element for the operation of a hospital facility, so it must be measured and managed both under technical and economic aspects. In recent years, the problems related to electrical energy consumption have achieved considerable importance. The procurement and proper use of electric energy are crucial steps for any complex structure wanting to reach an optimal level of energy management.

This work proposes a day-ahead load forecasting procedure. The load forecasting has been performed through the implementation of an ANN and has been tested to predict the electrical consumption of the Cellini clinic of Turin, which is a multispecialist hospital facility, part of the Humanitas Mirasole group. This type of approach has been chosen because it does not require a physical-mathematical model, representing the power consumption of the examined facility, which would have been much more complex to produce and thus not economical. In particular, the focus of this work is to provide a detailed analysis and an innovative formal procedure for the selection of all relevant ANN parameters (architecture type, inputs, activation functions, training algorithm, number of neurons) for modeling and predicting electrical loads in buildings.

9.2 Case study description

The Cellini clinic started its service in 1903 and received the accreditation by the National Health Service of Italy in 2003, because of structure and plant expansion. This has allowed the

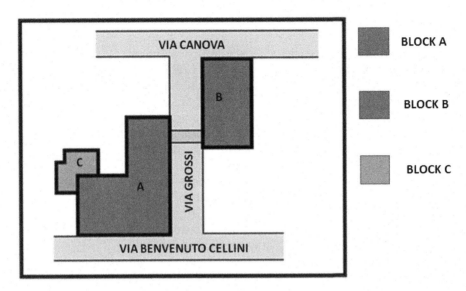

Fig. 9.1 Cellini clinic buildings.

Cellini clinic to double its space, enhancing the quality of its services with high technological and architectural standards.

The activities and the services offered by the Cellini clinic are considerable, given its structural and organizational size. During the year 2011 the clinic has performed 9698 hospital admissions according to the data of the National Health Agency, with a daily average of 27 recoveries. The clinic is spread over a total area of 9500 squares meters and is divided into three main blocks named A, B, and C (see Fig. 9.1).

The supply of electrical energy takes place through a medium voltage (15 kV) Point of Delivery (PoD). This PoD feeds two medium voltage/low voltage substations, named Station A and Station B, which present, respectively, one and two 630 kVA transformers. Station A supplies the block A of the Cellini clinic, while Station B feeds buildings B and C. Station A and Station B feed all the hospital equipment at low voltage. The whole clinic consumes nearly 3 million kWh a year of electric energy.

9.3 Dataset

In order to accurately evaluate the load demand of the Cellini clinic, the electrical energy consumption data, measured every 15 min, was gathered from the local distribution system operator. The collected data in the year 2012 amounts to 2662.325 MWh or 131.06 kWh/m^2.

From the available dataset it is possible to note that the energy consumption in summer is 30% higher than in winter due to air-conditioning. The winter load peaks are between 450 and 480 kW, while the summer peaks are between 600 and 650 kW.

The dataset also contains the ambient temperature that correlates with electrical consumption as presented in Table 9.1 (Matlab Manual, n.d.). In this case, the correlated vectors are the daily electrical energy consumption and the minimum/maximum temperature of the day. At the beginning of this study we considered the humidity, but removed it in the final dataset composition because of its low correlation with the load (see Table 9.1, where the correlation has been evaluated as in the temperature case).

The dataset was partitioned into a training set and a testing set. The training set was used to train an ANN and consists of an input vector and a score vector. The testing set was used to assess the strength and effectiveness of the predictive relationship and to calculate the forecasting error in terms of Mean Absolute Percentage Error (MAPE), Daily Peak MAPE, Coefficient of Variance of the Root Mean Squared Error (CVRMSE), Maximum Percentage Error (MPE), and Percentage of Test set with a MAPE under 5% ($PE_{5\%}$). These measures of accuracy are defined as follows:

$$\text{MAPE} = \frac{1}{N} \sum_{t=1}^{N} \frac{|Y_t - F_t|}{Y_t} \times 100\%$$

$$\text{Daily Peak MAPE} = \frac{\left| \max_t Y_t - \max_t F_t \right|}{\max_t Y_t}$$

Table 9.1 Correlation values between electrical consumption and minimum and maximum temperature as well as humidity.

	Minimum temperature	Maximum temperature	Minimum humidity	Maximum humidity
Correlation year 2012	0.72	0.70	0.10	0.12
Correlation Summer 2012	0.45	0.41	0.03	0.04
Correlation Winter 2012	0.43	0.32	0.02	0.03

$$CVRMSE = \frac{\sqrt{\dfrac{\sum_{t=1}^{N}(Y_t - F_t)^2}{N}}}{\overline{Y}} \times 100\%$$

$$MPE = \max_{t}\frac{|Y_t - F_t|}{Y_t} \times 100\% \; t = 1,\dots,N$$

where N is the number of data points, Y_t is the actual load, F_t is the predicted load, and $\overline{Y}$ is the mean of the testing set loads.

The Daily Peak MAPE analyzes the results with respect to daily load peaks, while the CVRMSE indicates the uncertainty in the forecast procedure. This last performance index has been exploited by the American Society of Heating, Refrigerating and Air-Conditioning Engineers (ASHRAE) to define a target for general model accuracy. ASHRAE recommends a CVRMSE value under 30% on hourly basis (Monfet et al., 2014).

9.4 ANN architecture

We propose an ANN consisting of an input layer, one hidden layer, and an output layer, as shown in Fig. 9.2. We chose to use only one hidden layer in order to have a good computational speed, considering that choosing all the ANN parameters with the procedure described in this chapter, several networks have to be trained many times on yearly dataset. In addition, calculation speed assumes huge importance for a future application in a BMS or a real-time monitoring system. The effectiveness of this choice was confirmed in all the simulation scenarios.

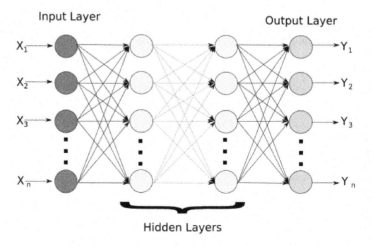

Fig. 9.2 General multilayer perceptron model.

One of the most important factors for designing a good ANN architecture is choosing appropriate input variables. In this chapter, two different types of ANN inputs are proposed.

Case A—No weather data. In this first dataset scenario, the weather data are not considered. The ANN inputs are as follows:

- Counter from 1 to 96 representing the 15-min intervals of the day.
- The day of the week from 0=Sunday to 6=Saturday.
- The 24-h-ahead average load.
- The day-ahead load.
- The 7-day-ahead load.
- Whether it is a holiday or weekday (0=holiday, 1=weekday).

Case B—Day-ahead temperature included. In addition to all the ANN inputs of Case A, the day-ahead temperature for the city of Turin is added. An example of this kind of input data is represented in Table 9.2.

After defining the ANN inputs, the layer transfer functions (activation functions) are analyzed. After some brief preliminary tests, four configurations were selected:

1. logsig/tansig: the Log Sigmoid Tangent Function ($logsig(x) = 1/(1 + \exp(-x))$) is used between input and hidden layer, while the Hyperbolic Tangent Sigmoid Transfer Function ($tansig(x) = 2/(1 + \exp(-2x)) - 1$) is between hidden and output layer.
2. tansig/purelin: tansig is used between the input and the hidden layer, while the Linear Transfer Function ($purelin(x) = x$) is used between the hidden and the output layer.

Table 9.2 ANN inputs (Case B) for the load forecast of Wednesday, June 6, 2012.

15 min intervals of the day	Day of week	24 h average load (kWh)	Day-ahead load (kWh)	7-day-ahead load (kWh)	Weekday holiday	Day-ahead temp. (°C)
1	3	77.25	73.48	61.60	1	21
2	3	77.27	80.52	65.92	1	21
3	3	77.31	82.41	66.64	1	21
⋮	⋮	⋮	⋮	⋮	⋮	⋮
96	3	85.18	67.32	63.48	1	18

3. logsig/purelin: logsig is used between the input and the hidden layer, while purelin is used between the hidden and the output layer.
4. tansig/tansig: tansig is used between all the layers.

Note that the purelin/purelin configuration has not been considered because preliminary tests were very unsatisfactory and overall, the relation between the selected ANN inputs and the future electrical energy consumption is not linear.

To decide which configuration is the best, four ANNs (one for each input scenario), with the four different transfer configurations, have been compared. These ANNs have been trained with the Resilient Backpropagation algorithm (RPROP), which will be analyzed hereinafter. The number of neurons in the hidden layer was fixed to 20.

The initialization of the biases and weights for the first ANN has been set randomly and replicated for the other ANNs as well as the selection of the number of neurons, and in all the final scenarios. The yearly datasets (Case A and Case B) were divided into 8 months of training and 4 months of testing data. Five tests have been executed for each input case. The results are reported in Tables 9.3 and 9.4.

The tansig/tansig configuration was chosen as the mean MAPE value is lowest. Note that the tansig function is mathematically equivalent to Hyperbolic Tangent ($y = \tanh x$). It differs in that it runs faster than the classic implementation of $\tanh x$. This choice is a good trade-off for a neural network, between speed, which is important mainly in the training phase, and the exact shape of the transfer function.

After defining the neural network inputs and the transfer functions, possible algorithms for training the neural network

Table 9.3 Testing results with different types of transfer configuration—Case A.

Test number	logsig/tansig MAPE (%)	tansig/purelin MAPE (%)	logsig/purelin MAPE (%)	tansig/tansig MAPE (%)
1	8.07	7.81	8.07	7.86
2	7.98	7.99	7.90	7.96
3	8.82	8.14	8.08	7.87
4	8.01	7.98	7.99	7.77
5	7.85	8.09	8.05	7.92
Mean	8.14	8.00	8.01	7.87

Table 9.4 Testing results with different types of transfer configuration—Case B.

Test number	logsig/tansig MAPE (%)	tansig/purelin MAPE (%)	logsig/purelin MAPE (%)	tansig/tansig MAPE (%)
1	8.39	8.28	8.29	8.16
2	8.28	8.18	8.28	8.20
3	8.42	8.22	8.35	8.23
4	8.51	8.31	8.41	8.11
5	8.36	8.32	8.21	8.30
Mean	8.39	8.26	8.30	8.20

have been analyzed to approach the best forecast accuracy. The training phase is crucial since the training algorithm tries to correlate the input data (historical energy consumption, day of the week, temperature, etc.) with the output data (real energy consumption).

In this work the Resilient Propagation algorithm (RPROP) was chosen, since it is not very sensitive to the number of neurons in the hidden layer (Bagnasco et al., 2014), the choice of which will be analyzed hereafter.

The slope of sigmoid functions approaches zero as the ANN inputs increase causing a problem when sigmoid functions and gradient descent are used, respectively, as transfer functions and training algorithm of a multilayer neural network. In this configuration the gradient can be characterized by a magnitude very near to zero and, therefore, cause small changes in the weights and biases, although these are distant from their optimal values. The RPROP algorithm only was the sign of the derivative provide the update direction of the weight; the derivative magnitude does not participate to the weight update. It is important to note that training is stopped if one of the following conditions holds:

- The number of iterations is greater than 1000.
- The gradient value is lower than 10^{-5}.
- The number of consecutive iterations wherein the gradient increases is greater than 5.

To complete the final structure of the ANN architecture, the number of neurons in the hidden layer has to be chosen. There is still no mathematical method to determine the number of neurons for hidden layers; it is usually found by using heuristic methods (Huang and Huang, 1991).

Table 9.5 Testing results for the neurons selection in the different input scenarios.

Case A		Case B	
No. of neurons	*MAPE (%)*	*No. of neurons*	*MAPE (%)*
6	8.47	6	8.31
8	7.78	6	7.75
10	7.87	10	7.68 ←
12	7.68	12	7.97
14	7.64	14	7.89
15	7.75	15	7.99
20	7.84	20	8.27
23	7.85	23	8.12
26	7.72	26	8.05
30	7.52 ←	30	8.08
32	7.59	32	7.98
33	7.60	33	7.95

The number of neurons was chosen according to MAPE. In this phase, tests have been done between 1 and 100 neurons with step equal to 1, see Table 9.5. As it can be seen, the best value for Case A is 30 and for Case B is 10.

The final ANN architectures, obtained after all the analysis described in this paragraph, are summarized in Fig. 9.3. Such an ANN forecasting tool can be integrated into real-time monitoring systems suitable for providing guidelines for energy efficiency and energy saving purposes (Bertolini et al., 2010; Bagnasco et al., 2013).

9.5 Results of simulation

The proposed ANN architectures have been tested to predict the electrical loads of the Cellini clinic. All the simulations performed for this work have been executed using Matlab (Matlab Manual, n.d.). To reduce the forecast errors seen in the previous paragraph, the yearly dataset has been divided into the four quarters of the year.

This subdivision was performed because of the differences in the load shape for every season. After this partitioning procedure, the ANN was trained and tested separately for each of these seasonal subsets and for each of the described input scenarios. Each seasonal dataset was divided into training and test sets.

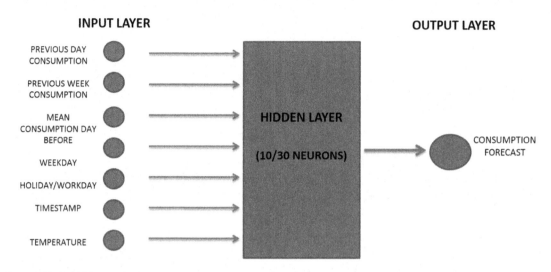

Fig. 9.3 Final ANNs architecture.

The training set was formed by the first two and half months of each quarter.

Tables 8–11 report the forecast errors in terms of MAPE, Daily Peak MAPE, CVRMSE, MPE, and $PE_{5\%}$ as well as the result of the *Lilliefors Test*, which establishes if the forecast errors are normally distributed (Lilliefors, 2012).

Figs. 9.4 and 9.6 show the comparison between the real and forecasted loads in one random week of the testing dataset. In both figures, the gray curve represents the real load while the black curve represents the forecasted load.

Figs. 9.5 and 9.7 report the statistical analysis of the prior plot's forecast error. In particular, each figure is composed of two subplots. The left one depicts the histogram of the forecast errors (gray bars) and a normal distribution fitted to the histogram

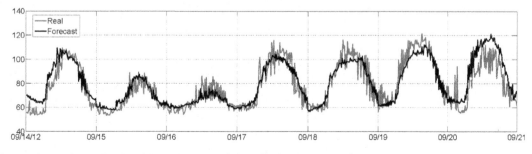

Fig. 9.4 Comparison of real and forecasted load (14–20 September, Case A).

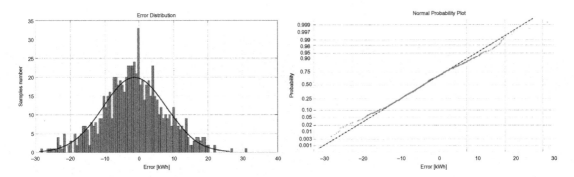

Fig. 9.5 Statistical analysis of the error (14–20 September, Case A).

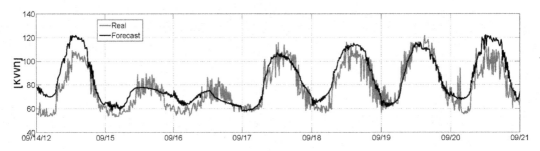

Fig. 9.6 Comparison of real and forecasted load (14–20 September, Case B).

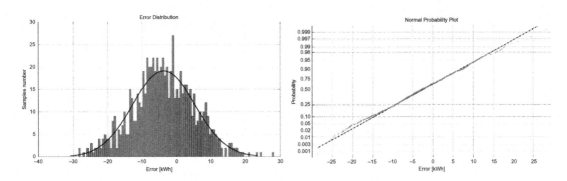

Fig. 9.7 Statistical analysis of the error (14–20 September, Case B).

(black line). The right one shows a normal probability plot of the forecast error with the errors displayed as a set of "+" symbols. If the errors are distributed normally, or according to a Gaussian distribution, the plot will be nearly linear. The goal of these two plots is to assess the normality of the forecast error graphically.

Table 9.6 Forecast errors for the 15 days of March 2012.

		MAPE (%)	Daily peak MAPE (%)	CVRMSE (%)	MPE (%)	PE$_{5\%}$ (%)	Lilliefors test
March	Case A	6.47	14.85	8.26	24.4	58.67	X
	Case B	6.75	16.44	8.55	27.7	58.11	X
July	Case A	6.78	10.72	8.68	30.47	52.99	✓
	Case B	7.42	11.12	9.39	34.21	48.30	✓
September	Case A	7.79	7.80	10.94	42.33	49.77	✓
	Case B	8.97	7.86	11.15	39.27	46.12	✓
December	Case A	5.64	3.84	6.97	24.37	68.98	✓
	Case B	5.70	3.93	7.23	26.17	64.19	✓

From Table 9.6, we observe a mean MAPE close to 7%. Since the load curves are not smooth, and the ANNs have been trained on a small dataset (about 7200 values, for each subset), we can conclude that a mean forecasting error of about 7% is an excellent result, which demonstrates the reliability and robustness of the proposed prediction strategy. As shown previously, the forecasts are more accurate during the winter months, where the mean MAPE is less than 6%. This is because the winter energy consumption is lowest and there is no air-conditioning.

The daily peak MAPE, the mean value of which is around 9.5%, indicates a worse forecasting capability of daily peak loads, due to the irregular presence of peaks. The temperature, included in the Case B dataset, does not seem to be necessary, since it does not lead to a prediction improvement. This is because the correlation between electrical load and temperature does not appear to be high in the case of a hospital. To improve the algorithm performance in the critical months, i.e., August and September, it is possible to execute a monthly training (Shahidehpour et al., 2002). This enhancement can be highly effective if the dataset contains those particular months for at least three different years.

The forecast errors in terms of MAPE are similar to recent works related to load forecasting for buildings. In Escriva-Escriva et al. (2011) and Roldan-Blay et al. (2013), a neural network using an hourly temperature curve model is tested to forecast the consumption of the Universitat Politecnica de Valencia. They report a MAPE close to 7% most of their simulation scenarios. In Li and Chu (2011), a neural network and hybrid neuro-fuzzy system are exploited to predict buildings consumption with result comparable to those presented in Table 9.6.

The forecast errors in terms of CVRMSE are similar to recent works related to load forecasting. In particular, considering the longer time horizon of the proposed procedure, the results seem to be improved compared to Monfet et al. (2014) and Platon et al. (2015). In addition, all CVRMSE values result to be clearly lower than the ASHRAE limits.

MPE values can be useful to detect anomalous energy consumption. For example, it can be noted from Figs. 9.4 and 9.6 that there was an irregular energy absorption during the night of the 20th September 2012, which leads to a MPE of 42.33% in Case A and 39.27% in Case B. Including a control on the MPE value in the monitoring system, this type of anomalous consumption could be easily found. Furthermore, the MPE values in the seasonal subsets 1, 2, and 4 are less than 30%.

The PE5% mean value is around 57%, which means that almost 60% of the 15-min intervals of the day are characterized by a MAPE value lower than 5%. The statistical results are very satisfactory since Figs. 9.5 and 9.7, related to the statistical analysis of the forecast error, show a good approximation to the normal distribution. In addition, the Lilliefors test has a negative response only in the first seasonal subset (January-February-March), as reported in Table 9.6. In this case, the negative answer to the Lilliefors test is due to the worse value of the Daily Peak MAPE caused by an irregular and certainly not optimal behavior of the energy consumption.

9.6　Conclusions

A multilayer perceptron ANN, based on a backpropagation algorithm, is proposed for forecasting the day-ahead load consumption of a large hospital facility. In addition, a detailed analysis, and an innovative formal procedure for the selection of all ANN parameters (architecture type, inputs, transfer functions, training algorithm, number of neurons) was described.

Results proved to be satisfactory, with reasonable errors in terms of MAPE, CVRMSE (always under ASHRAE limits), and statistical behavior of the forecast error. The good performance achieved (MAPE and PE5% mean value, respectively, close to 7% and 60%) suggests that a similar approach could be applied to forecast other types of loads (i.e., domestic, industrial).

Electrical load forecasting and modeling is an essential component for a better dynamic management of thermal behavior of buildings (HVAC) for energy saving policy. In fact, the presented forecast algorithm could also be easily integrated into a BMS real-time monitoring system.

9.7 Practical utilization

The resulting model was deployed in the Building Management System (BMS) of Clinica Cellini and put into real-time operation. It has been in use there since 2010. The overall electricity consumption at Humanitas Cellini has been reduced from 3100 to 2100 MWh (32%) per year from 2010 to 2019. In the same timeframe, the sister hospital Humanitas Gradenigo reduced its demand from 5100 to 3700 MWh (27%) per year.

These reductions are partially due to engineering changes like HVAC maintenance, replacing lightbulbs with LED lights, and retrofitting refrigeration and heat exchanger systems. A major aspect was the creating of air handling unit scheduling and thermal compensation curves. The proper scheduling of air-conditioning can flatten the curves visible in Figs. 9.4 and 9.6. Scheduling these HVAC activities is only possible if there is an accurate and effective forward-looking system as presented here.

The biggest contribution was the behavioral program that convinced staff and patients to be content at different temperatures than before. This was made possible by the forecasting system that allows people to look ahead at scenarios in the future and make choices about them. Being made aware of future repercussions of today's choices enabled objective decisions in favor of less electricity use.

Changing peoples' minds is hard work. Change management is a slow process and must be taken seriously whenever many people are expected to do something differently that they are used to. Good early results and the experience of retained comfort at different temperatures helped to convince them to follow along and adopt the system.

In conclusion, a load forecasting system as discussed here combined with change management can lead to a reduction of a building's electricity consumption by one-third.

References

Abdel-Aal, R.E., 2004. Short-term load forecasting using abductive networks. IEEE Trans. Power Syst. 19 (1), 164–173.

Amjady, N., Keyna, F., 2009. Short-term load forecasting of power system by combination of wavelet transform and neuroevolutionary algorithm. Energy 34, 46–57.

Amjady, N., Keyna, F., Zareipour, H., 2010. Short-term load forecast of microgrids by a new bi-level prediction strategy. IEEE Trans. Smart Grid 1 (3), 286–294.

Bagnasco, A., Massucco, S., Silvestro, F., Vinci, A., 2013. Monitoraggio Intelligente dei Consumi dell'Ateneo Genovese. In: Automazione e strumentazione, Marzo 2013, n. 2, pp. 82–85.

Bagnasco, A., Saviozzi, M., Silvestro, F., Vinci, A., Grillo, S., Zennaro, E., 2014. Artificial neural network application to load forecasting in a large hospital facility. In: 13th International Conference on Probabilistic Methods Applied to Power Systems (PMAPS), Durham, 7-10 July.

Bertolini, S., Giacomini, M., Grillo, S., Massucco, S., Silvestro, F., 2010. Coordinated micro-generation and load management for energy saving policies. In: Innovative Smart Grid Technologies Conference Europe (ISGT Europe).

Ceperic, E., Ceperic, V., Baric, A., 2013. A strategy for short-term load forecasting by support vector regression machines. IEEE Trans. Power Syst. 28 (4), 4356–4364.

Doukas, H., Patlitzianas, K.D., Iatropoulos, K., Psarras, J., 2007. Intelligent building energy management system using rule sets. Build. Environ. 42, 3562–3569.

Escriva-Escriva, G., Alvarez-Bel, C., Roldan-Blay, C., Alcazar-Ortega, M., 2011. New artificial neural network prediction method for electrical consumption forecasting based on building end-users. Energy Build. 43, 3113–3119.

Grillo, S., Marinelli, M., Massucco, S., Silvestro, F., 2012. Optimal management strategy of a battery-based storage system to improve renewable energy integration in distribution networks. IEEE Trans. Smart Grid 3 (2), 950–958.

Hong, T., Wilson, J., Xie, J.I., 2014. Long term probabilistic load forecasting and normalization with hourly information. IEEE Trans. Smart Grid 5 (1), 456–462.

Huang, S.-C., Huang, Y.-F., 1991. Bounds on the number of hidden neurons in multilayer perceptrons. IEEE Trans. Neural Netw. 2 (1), 47–55.

Jetcheva, J.G., Majidpour, M., Chen, W.P., 2014. Neural network model ensembles for building-level electricity. Energy Build. 84, 214–223.

Khosravi, A., Nahavandi, S., 2014. Load forecasting using interval type-2 fuzzy logic systems: optimal type reduction. IEEE Trans. Ind. Inf. 10 (2), 1055–1063.

Krstic, H., Koski, Z., Otkovic, I.S., Spanic, M., 2014. Application of neural networks in predictiong airtightness of residential units. Energy Build. 84, 160–168.

Li, H.S., Chu, J., 2011. Forecasting building energy consumption using neural networks and hybrid neuro-fuzzy system: a comparative study. Energy Build. 43, 2893–2899.

Lilliefors, H., 2012. On the Kolmogorov Smirnov test for normality with mean and variance unknown. J. Am. Stat. Assoc. 62, 399–402.

Lu, Y., Shi, H.-F., 2012. The hourly load forecasting based on linear Gaussian state space model. In: Machine Learning and Cybernetics (ICMLC), 2012 International Conference on, vol. 2, 15-17 July, pp. 741–747.

Matlab Manual. www.mathworks.com.

Missaoui, R., Joumaa, H., Ploix, S., Bacha, S., 2014. Managing energy smart homes according to energy prices: analysis of a building energy management system. Energy Build. 71, 155–167.

Monfet, D., Corsi, M., Choiniere, D., Arkhipova, E., 2014. Development of an energy prediction tool for commercial buildings using case-based reasoning. Energy Build. 81, 152–160.

OECD/IEA, International Energy Agency, 2014. http://www.iea.org/aboutus/faqs/energyefficiency/.

Platon, R., Dehkordi, V.R., Martel, J., 2015. Hourly prediction of a building's electricity consumption using case-based reasoning, artificial neural networks and principal component analysis. Energy Build. 92, 10–18.

Roldan-Blay, C., Escriva-Escriva, G., Alvarez-Bel, C., Roldan-Porta, C., Rodriguez-Garcia, J., 2013. Upgrade of an artificial neural network prediction method for electrical consumption forecasting using an hourly temperature curve model. Energy Build. 60, 38–46.

RTE: rseau de transport d'lectricit, Le bilan lectrique francais 2010. http://www.rte-france.com/.

Shahidehpour, M., Yamin, H., Li, Z., 2002. Market Operation in Electric Power Systems. John Wiley & Sons.

Webberley, A., Gao, D.W., 2013. Study of artificial neural network based short term load forecasting. In: Power and Energy Society General Meeting (PES), 2013 IEEE, 21–25 July, pp. 1–4.

Yan, J., Tian, C., Huang, J., Wang, Y., 2012. Load forecasting using twin Gaussian process model. In: Service Operations and Logistics, and Informatics (SOLI), 2012 IEEE International Conference on 8-10 July, pp. 36–41.

Ying, C., Luh, P.B., Che, G., Yige, Z., Michel, L.D., Coolbeth, M.A., Friedland, P., Rourke, S.J., 2010. Short-term load forecasting: similar day-based wavelet neural networks. IEEE Trans. Power Syst. 25 (1), 322–330.

Further reading

Nocedal, J., Wright, S.J., 1999. Numerical Optimization. Springer.

Soft sensors for NOx emissions

Patrick Bangert[a,b]

[a]*Artificial Intelligence, Samsung SDSA, San Jose, CA, United States*
[b]*Algorithmica Technologies GmbH, Bad Nauheim, Germany*

10.1 Introduction to soft sensing

A **sensor** is a physical device that measures a quantity due to a physical reaction to a change in the environment of the device. For example, a temperature sensor usually consists of a material called a thermistor that changes its electrical resistance in response to changes in temperature. The electrical resistance is measured using ohmmeter. During the calibration phase, a researcher would establish a conversion formula that can convert ohmmeter readings into temperature readings. This formula is usually a simple relationship, but it is instructive to note that even simple and regular sensors require a model to convert the physical effect of note (here the resistance) to the effect being measured (here the temperature).

A **soft sensor** is a formula that converts various inputs from simple sensors and combines them to mimic the output of a more complex sensor. In this chapter, we will discuss a sensor for the nitrogen oxides (NOx) emissions of a cogeneration power plant. Directly measuring the NOx is possible but the sensor is expensive

Machine Learning and Data Science in the Power Generation Industry. https://doi.org/10.1016/B978-0-12-819742-4.00010-X
© 2021 Elsevier Inc. All rights reserved.

and fragile. If it were possible, which it is, to substitute it with a formula that is based on a set of much cheaper and robust sensors (such as temperatures, pressures, flowrates, and so on), then we would have several benefits from this:

1. The value from the soft sensor is cheaper both initially and over the lifetime of the plant as no device needs to be purchased or maintained.
2. It is always available as the soft sensor will never fail or need to be removed for recalibration.
3. It is available in real time as it is just a calculation and we do not have to wait for some physical reaction to take place.
4. It is scalable over many assets or locations without investment or difficulty.

It makes sense to substitute expensive and fragile sensors with soft sensors. Many plants must rely on laboratories to perform certain measurements that are too difficult to perform in the stream of the plant. This necessitates the taking of a sample that is usually done by a person. The sample is taken to the laboratory that measures the value in question. The values are usually made available by manually typing the value into a spreadsheet, which becomes available several hours after the sample was originally taken. In this way, we can expect to receive one or two values per day, at most. This process is expensive, slow, and error prone. A soft sensor can eradicate the cumbersome nature of this process and make a continuous value available to the plant. In turn, this value then can be used in advanced process control applications that can improve the workings of the plant. Therefore a soft sensor can directly contribute to the bottom line of a plant by enabling hitherto impossible control strategies.

Measuring pollutants like NOx and SOx, gas chromatography in chemical applications, or multiphase flow in upstream oil and gas applications are some examples where this has been successfully done on an industrial scale.

The idea of a soft sensor is the same as the idea of the thermistor temperature sensor. We are measuring something simple such as the electrical resistance that we then convert computationally into what we really want to know such as the temperature. The only two differences are that we now have multiple simple quantities that flow into the computation and, usually, a nonlinear relationship between these quantities to get to the final output. These form our dual challenge.

First, we must collect as much domain knowledge about the system as we can to determine the fundamental question: Which simple measurements do we need in order to be able to compute the quantity of interest? If we leave out something important, the quality of the model may be poor. If we include something that is not connected to the output, then this may disturb the calculation.

This idea is subtle and so let us look at an everyday example of this. The sale of ice cream is correlated to the number of drownings in swimming pools. While this is true, the relationship is obviously not causal. It is the higher ambient temperature that results in higher ice cream sales and more pools visits, which in turn increase the drownings. It would be better to predict the ice cream sales using temperature and not the number of drownings. This is obvious to a human being who knows what is going on—the domain expert—and not at all obvious to an automated computer analysis method that just looks at correlations.

Second, we must establish the nonlinear relationship between all the variables to get the best formula to calculate our desired output. This is the realm of machine learning.

10.2 NOx and SOx emissions

NOx is an umbrella term combining several different nitrogen oxides. Most notably among them are nitric oxide (NO) and nitrogen dioxide (NO_2). These gases contribute to smog, acid rain, and have various detrimental health effects. The term also includes nitrous oxide (N_2O), which is a major greenhouse gas.

SOx is an umbrella term combining several different sulfur oxides. Most notably among these are sulfur dioxide (SO_2), which is a toxic gas, and sulfur trioxide (SO_3), which is the main component of acid rain. The combination of NO_2 with SOx will form sulfuric acid, which is harmful.

VOC is an umbrella term for volatile organic compounds, which are a variety of chemicals that have detrimental health effects.

The combustion of fossil fuels produces all these gases. They must be removed by so-called scrubbers which spray a reactant into a chamber with the gas that reacts with the unwanted chemicals and converts them into a chemical that is harmless and can be captured. For example, SO_2 is converted by limestone into gypsum, which is a material used by the construction and agricultural industries. While most of the produced pollutants is captured, some amount is released into the atmosphere causing several unwanted effects such as the greenhouse effect. How much of each gas is released is the question at hand.

The recognition that these gasses are harmful has led to governmental regulations capping the amount that may be released.

In the case of another greenhouse gas, CO_2, it has even led to the creation of a carbon credit economy where permissions to release the gas are traded on the financial markets. There is significant interest on the part of the general public, governments, and regulators to keep the released gasses to a minimum. The amount of these gasses that are produced by combusting a specific amount of the fossil fuel is determined by the fuel itself—even though it varies significantly depending on where the fossil fuel was mined—the critical element in this effort is the scrubber system. Knowing exactly how much is being released and knowing it in real time allows advanced process control in order to properly operate the scrubber and therefore minimize the emitted gasses.

The regulators require the emission to be determined with accompanying documentation to make sure that permitted limits are adhered to (California Energy Commission, 2015). The standard method of doing this is to measure these gases with a sensor array and to record the values in a data historian.

Three problems are encountered in practice with this approach. First, the sensor array is usually quite expensive running into several hundred thousand dollars for each installation. Second, the sensor array is fragile especially in the harsh environments typical of the process industry, which in turn leads to significant maintenance costs and monitoring efforts that distract from the process equipment. Third, whenever the sensor array does not provide values—due to a failure or temporary problem—the regulator may conclude that permitted limits have just been exceeded and charge a fine. Fines are costly but may even lead to limits being reduced or public perception being affected.

If the released amounts could be calculated from process measurements, rather than measured directly, then this would solve all the earlier discussed problems. A calculation is cheap, robust, and does not break down.

Due to all these factors, the soft sensing of emissions is directly tied to the minimization of emissions and the adherence to regulations.

10.3 Combined heat and power

In our application, we consider the process of a **combined heat and power** (CHP) generation station, also known as **cogeneration**. This type of power plant uses a heat engine, such as a gas turbine, to generate both electricity and heat. The combustion of fossil fuels produces high-temperature gas that drives the turbine to produce electricity and the low-temperature waste heat is then provided as the heat output, often in the form of steam.

In applications where both electricity and heat are needed, cogeneration is significantly more efficient (about 90%) than generating electricity and heat separately (about 55%). As such, this is a more environmentally friendly way to generate power.

See Fig. 10.1 for an overview of the cogeneration setup. At the beginning air is mixed with natural gas to be combusted in a gas turbine. The hot gas causes the turbine to rotate, which in turn rotates the generator that makes electricity. The exhaust gas is treated in the heat recovery steam generator (HRSG). This uses the hot gas to heat water in order to generate high-pressure and high-temperature steam. The steam is passed through a steam turbine connected to a second generator that produces further electricity that would have been lost if we had not recovered it. The remaining steam at this point can be extracted to provide heat to some application. The rest is cooled and can be recycled in the steam cycle.

As a last step, just before the stack, the flue gas is scrubbed in a scrubber, see Fig. 10.2. The flue gas enters at the bottom and the scrubbing liquid (for example pulverized limestone in water) enters at the top and flows down through various levels. The levels provide the gas multiple opportunities to react with the liquid before it exits at the top. Such scrubbers generally remove about 95% of the pollutants in the gas. For extra purity, a second scrubber can be added.

10.4 Soft sensing and machine learning

A soft sensor is a formula that calculates a quantity, y say, from measured quantities, $\underline{x}$ say, so that the quantity of interest does not have to be measured. This formula $y = f(x)$ can take any form. In many cases, we have an explicit formula that is known from physics or chemistry. In some cases, however, we do not have a known formula and so the formula must be determined empirically from available data.

Determining a formula from data is the purpose of machine learning (Goodfellow et al., 2016). Usually the function takes one of a few standard forms such as a linear regression (Hastie et al., 2016), feed-forward neural network (Hagan et al., 2002), recurrent neural network (Yu et al., 2019), random forest (Parmar et al., 2018), and others (Bangert, 2012). The choice of the form is made by the data scientist and machine learning determines the parameters that make the chosen form fit best to the data. Having gotten the best parameters, we can try the formula on some more empirical data to see how it performs. If the accuracy of the formula on this new data is better than the

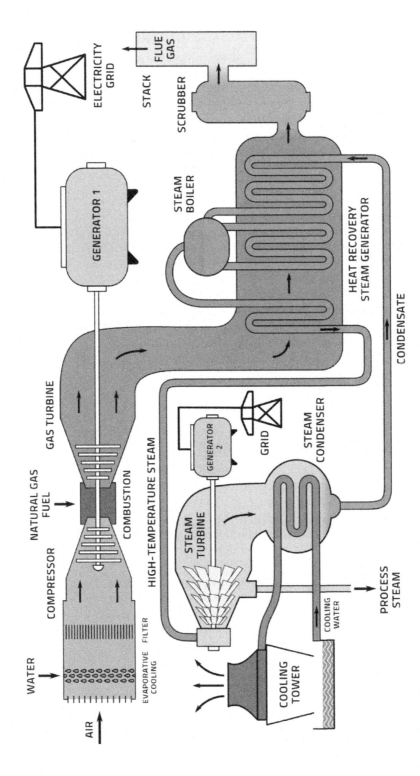

Fig. 10.1 The general setup of a cogeneration plant.

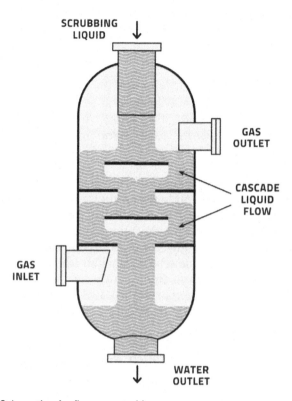

Fig. 10.2 Schematic of a flue gas scrubber.

required threshold, then the soft sensor is finished. In this sense, machine learning a soft sensor is comparable to calibrating a physical sensor.

As the physical sensor is the standard, we must compare any soft sensor to it on the same principles that a physical sensor is deemed fit for its purpose. This is **calibration**. According to the International Bureau of Weights and Measures (BIPM), calibration is an "operation that, under specified conditions, in a first step, establishes a relation between the quantity values with measurement uncertainties provided by measurement standards and corresponding indications with associated measurement uncertainties (of the calibrated instrument or secondary standard) and, in a second step, uses this information to establish a relation for obtaining a measurement result from an indication" (JCGM, 2008).

In this sense, calibration is the process of comparing the output of the sensor in question to some standard. If the deviation between the two is less than a certain amount, then calibration is successful. The deviation must always be seen as relative to the measurement uncertainty of both the sensor in question as well as the standard to which it is being compared. The usual assessments of machine learning with its root-mean-square-errors and distribution of residuals are exactly this kind of assessment.

There are various standards around the world that require sensors to be calibrated and recalibrated at regular intervals. These are, for example, ISO-9001, ISO 17025, ANSI/NCSL Z540, and MIL-STD-45662A. However, these standards do not define the precise requirement that the sensor has to satisfy in order to be deemed calibrated.

Whether a sensor is calibrated or not is usually decided based on the maximum deviation between the sensor in question and the standard as compared to the measurement uncertainty of the sensor in question. If the deviation is always less than ¼ of the measurement uncertainty, the sensor is considered calibrated. This ratio of ¼ is, of course, a somewhat arbitrary choice that emerged historically due to the practices of the military of the United States (Jabloński and Březina, 2012; DOD, 1984). Since this is an emergent practice and not a legally defined requirement, it is often up to the regulator to define this ratio and it can be larger, such as ½. The other variable is the measurement uncertainty. If the sensor in question is outfitted with an appropriately large uncertainty, it will always meet these criteria. So, the question is not whether a sensor is calibrated or not but rather (a) at what ratio and (b) with what uncertainty the sensor is calibrated.

In conclusion, if we have a good model in the sense of machine learning, it is ipso facto a calibrated model. Providing this assessment and documentation to a regulator will allow them to accept a soft sensor in lieu of a physical sensor relative to the same standards.

10.5 Setting up a soft sensor

To begin with, all available empirical data is examined by domain experts to extract those sensors that have the most relevant information content to produce NOx or SOx. It is important to present all the information that we have and not to present any distracting data that may only serve to distract the model. Machine learning is fundamentally unable to differentiate between correlation and causation, which makes the initial selection of sensors vital to the success and this is best performed by an experienced expert (Guyon and Elisseeff, 2003).

The full dataset is now split into a training dataset that will be used to make the model and a testing dataset that will be used to assess the goodness of fit of the model. It is typical to randomly select 75% of the data for training and the rest for testing.

Two basic problems plague machine learning and they are underfitting and overfitting (Bishop, 2006). Underfitting results from having so few parameters that the model cannot possibly represent the phenomenon. Overfitting results from having so many parameters that the model can memorize the data without learning the underlying dynamics. Clearly, both possibilities will produce a poor model in the sense that a novel data point will obtain an unsatisfactory computational result. To avoid both, we must find a medium number of parameters and this is usually problem dependent. A very rough rule of thumb is to obtain at least 10 data points for every one parameter in the model. This is a rough rule as the information content is not proportional to the number of points in a dataset; for instance, if a point is duplicated, then a point has been added but no new information was added. A solution typically recommended in machine learning textbooks—getting more data—is often infeasible especially in an industrial context where data acquisition entails costs of money and time. The available data therefore puts an a priori limit on the possible complexity of the model.

In order to reduce the necessary parameter count in the model while keeping the information content in the dataset the same, we usually employ a dimensionality reduction method. Such a method takes the dataset and projects it into a space with fewer dimensions without merely deleting any dimensions. There are many such methods but the most popular is principal component analysis (Guyon and Elisseeff, 2003).

For the cogeneration plant that we are dealing with here, the following variables were selected as suitable inputs to a soft sensor:

1. Heat recovery steam generator (HRSG).
   a. Burner: temperatures, fuel pressure, flame strengths.
   b. Feedwater temperature rise (FWTR): temperature, pressure, flowrate.
   c. Control valves: opening.
   d. Ammonia: leak detector.
   e. General: inlet temperature, tube skin temperatures, steam pressure, flowrates, turbine exhaust heat rate, position feedback, heat rate, differential pressure across selective catalyst reduction (SCR), heat recovery, water flowrate, fuel flowrate, quality trim factor, consumed heat water.
2. Gas turbine.
   a. General: fuel pressures, fuel volume rates, fuel heat rates, ammonia mass injection.

b. Engine: thermocouple, variable frequency drive frequency, compensator temperature.

3. Weather.

a. General: wind direction, rain rate, air pressure, humidity.

As some of these variables are available multiple times, we had 31 variables available that are the vector x. They are sampled once every 10 min. The NOx and SOx output as measured by the sensor array is also provided and that is the known y that we are going to use to train our soft sensor using machine learning. We perform principal component analysis to reduce the dimensionality from 31 to 26. This retains 99% of the variance of the original dataset. A deep feed-forward neural network is trained on data from 1 year.

In order to demonstrate the model capability for diversity, we choose to model the NOx concentration in parts per million and the SOx flowrate in pounds per hour.

10.6   Assessing the model

In training the model, we split the data into three parts. The training data is used to adjust the model parameters so that the model is as close to these training data as possible. The testing data is used to evaluate when this training is making no more significant progress and can be considered finished. The validation is not used for training at all. The finished model is executed on the validation data so that we can judge how well the model performs on data it has never seen before. This is the true test of how well the model performs. This would be the basis for any calibration documentation provided to a regulator.

We thus compare the computed result of the model y' with the known result of the measurement, y, using all three datasets. For every data point x_i, we compute the residual $r_i = y_i - y'_i$. If the residual is very close to zero, the model is doing well and if it is far away from zero, the model is doing poorly. In order to study many data points it is helpful to plot the probability distribution of residuals, see Fig. 10.3.

This distribution is arrived at by counting how often a particular residual occurs over a large dataset. We then plot the residual on the horizontal axis and the frequency of occurrence on the vertical axis. Where the curve is highest is then the most often occurring residual.

We expect that this distribution is a bell-shaped curve, centered on zero, symmetrical about zero, exponentially decaying to either side, and having a small width. What makes the width "small" depends on the use case. Some uses need a more accurate

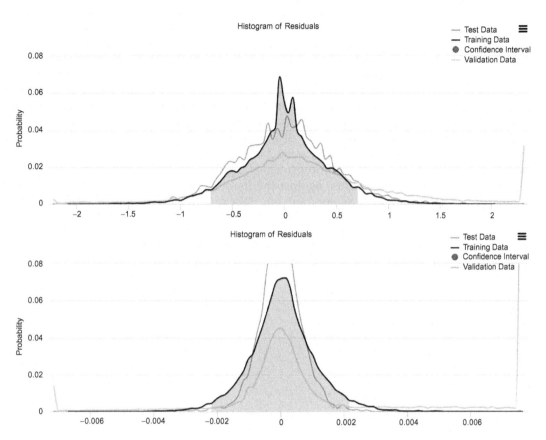

Fig. 10.3 The probability distribution of residuals of NOx and SOx for training data, testing data, and validation data. All three distributions are bell-shaped curves centered on zero and so are what we expect to see.

model than others and this is the most important success criterion to be set for the model. It is usually unrealistic to expect this distribution to be normal as the normal distribution usually decays faster in its tails than the distributions we encounter in practice. This is due to a variety of factors some of which are artifacts of empirical data analysis such as a limited number of data points. Other factors are due to the measurement process itself, i.e., there are causal factors that are not measured and cannot be included in the dataset at all (Tsai et al., 1998).

Loosely speaking, the standard deviation of the residual distribution is the accuracy we can expect of the model and it is the final conclusion to be presented upon calibration.

In our case, the distribution for the residuals for training, testing, and validation adhere to these principles. We see that the residuals are larger for testing and validation than for training.

This behavior is to be expected as a model usually performs better on data that was used to make the model. This is probably also due to the system slightly changing its behavior over time, i.e., aging. There are two factors to the aging process in industrial applications. First, there is genuine mechanical degradation over time of all manner of physical components of the system. Second, there is sensor drift that changes the numerical values recorded for any sensor. It is due to sensor drift that sensors occasionally need recalibration.

In addition to checking the plot of residuals, we may plot the measured against the modeled value directly, see Fig. 10.4. Here we can see that the data points are distributed closely around

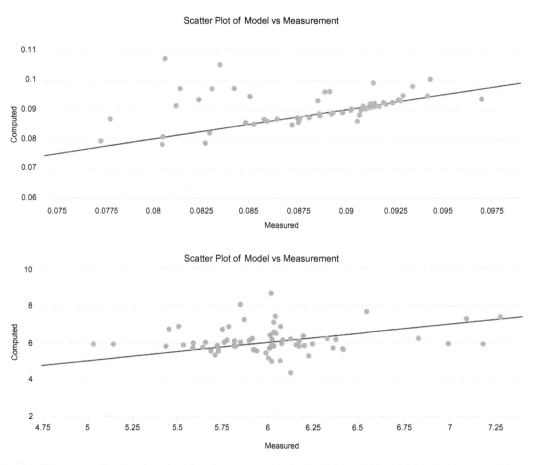

Fig. 10.4 The computed value plotted against the measured value for NOx and SOx. Ideally, this data would form a *straight line*, displayed in *black* for reference.

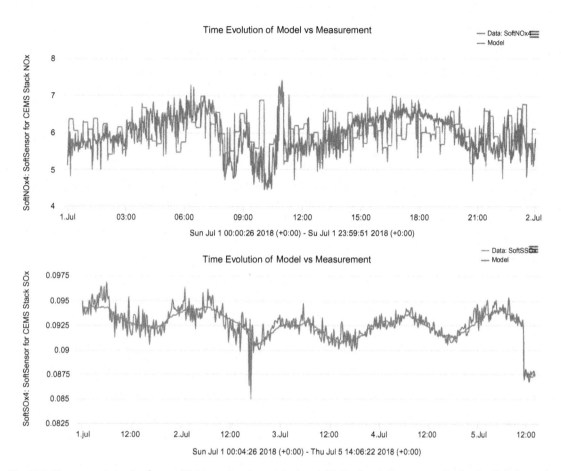

Fig. 10.5 The computed value [*green* (*light gray* in print version) variable line] and the measured value [*red* (*dark gray* in print version) smooth and continuous line] for NOx and SOx.

the ideal straight line where the computation would equal the measurement.

Finally, we may plot a timeline of the measurement and the computation to assess the true values against each other over time, see Fig. 10.5. This is the output that an operator would see.

In assessing these figures, we must keep in mind that no sensor is precise in its measurement. Every measurement is uncertain to some degree and has an inherent uncertainty. The inherent uncertainty is best assessed by the standard deviation of the variable. In our case, this is 2.43 ppmc for NOx and 0.02 lb/h for SOx. Comparing this to the standard deviation of 0.75 ppmc for NOx and 0.01 lb/h for SOx in the results of the physical measurement, we may conclude that the soft sensor is roughly as accurate as the physical sensor array.

10.7 Conclusion

We find that the model for NOx and SOx is comparable in accuracy to the physical measurement. This makes the models suitable for use in real CHP plants instead of the physical sensor array. This use saves significant cost in the acquisition and maintenance of the sensors as well as in the fines associated with sensor malfunction and failure.

As the models were made using machine learning based on empirical data, they did not consume much human time and effort. However, they benefitted significantly from human expertise mainly due to the careful selection of input data. Since the models include the important human expertise and that they were carefully examined against measurement data, we can conclude that the model output is reliable.

References

Bangert, P., 2012. Optimization for Industrial Problems. Springer, Heidelberg.

Bishop, C.M., 2006. Pattern Recognition and Machine Learning. Springer, Heidelberg.

California Energy Commission, 2015. Guidelines for certification of combined heat and power systems pursuant to the waste heat and carbon emissions reduction act. California Energy Commission CEC-200-2015-001-CMF.

Department of Defense, 1984. Military handbook: evaluation of contractor's calibration system. Department of Defense MIL-HDBK-52A.

Goodfellow, I., Bengio, Y., Courville, A., 2016. Deep Learning. MIT Press, Boston.

Guyon, I., Elisseeff, A., 2003. An introduction to variable and feature selection. J. Mach. Learn. Res. 3, 1157–1182.

Hagan, M.T., Demuth, H.B., Beale, M., 2002. Neural Network Design. PWS Publishing, Boston.

Hastie, T., Tibshirani, R., Friedman, J., 2016. The Elements of Statistical Learning: Data Mining, Inference, and Prediction, second ed. Springer, Heidelberg.

Jabloński, R., Březina, T., 2012. Mechatronics. Springer, Heidelberg.

JCGM, 2008. International Vocabulary of Metrology—Basic and General Concepts and Associated Terms (VIM). BIPM, Sèvres.

Parmar, A., Katariya, R., Patel, V., 2018. A review on random forest: an ensemble classifier. In: Proceedings of ICICI 2018: International Conference on Intelligent Data Communication Technologies and Internet of Things (ICICI). Springer, Heidelberg.

Tsai, C.L., Cai, Z., Wu, X., 1998. The examination of residual plots. Stat. Sin. 8, 445–465.

Yu, Y., Si, X., Hu, C., Zhang, J., 2019. A review of recurrent neural networks: LSTM cells and network architectures. Neural Comput. 31, 1235–1270.

11

Variable identification for power plant efficiency

Stewart Nicholson[a] and Patrick Bangert[b,c]

[a]*Primex Process Specialists, Warrington, PA, United States*
[b]*Artificial Intelligence, Samsung SDSA, San Jose, CA, United States*
[c]*Algorithmica Technologies GmbH, Bad Nauheim, Germany*

Chapter outline

11.1   Power plant efficiency

The efficiency of a power plant is a measure of how much of the potential energy contained in the fuel source is converted to usable energy provided to the consumer in the form of electricity, and in some cases heat. Every part of the process of converting a unit of fuel into power loses some of this energy. In this chapter, we will focus on fossil fuel power plants that are fundamentally based on combusting some fuel source such as coal, gas, or oil. Some of the energy losses are chemical because the fuel is converted to ash. Some losses are thermal because of nonperfect insulation. Some are frictional because moving parts like the turbine are not frictionless. Some losses are electrical because the power plant itself requires electricity to run.

Most of the energy loses are thermal however, which is the main reason why combined heat and power plants are so much more efficient as opposed to pure electricity-generating power plants. The heat left over after the electricity has been generated can be used to provide useful heat to industry or to homes. Typical efficiencies   are   about   37%   for   coal-   and   oil-fired

Machine Learning and Data Science in the Power Generation Industry. https://doi.org/10.1016/B978-0-12-819742-4.00011-1
© 2021 Elsevier Inc. All rights reserved.

electricity-generating plants and about 60% for combined heat and power plants. Even in the latter case, this means that about 40% of the potential energy in the fuel source are lost. Approximately half of these losses are due to heat escaping out of the stack and the other half are lost in the cooling water cycle.

The most common measure of efficiency in North America is the **heat rate**. It is the amount of energy used in order to generate 1 kWh of electricity. The amount of energy is measured in British thermal units (Btu). This non-SI unit has multiple definitions, but it is equal to between 1054 and 1060 kJ. According to the US Energy Information Administration, 3412 Btu is the energy content of 1 kWh of electricity. Therefore if the heat rate of a power plant is 10,500 Btu/kWh, the efficiency is 32.5%. Calculating the heat rate is not a triviality however as demonstrated in CEC (2019).

Another way to calculate the efficiency is the **Rankine cycle** view. The Rankine cycle is the basic template for a thermal power plant. Gas turbine efficiency is more accurately described by the related Brayton cycle. In most thermal power plants, fossil fuel is combusted to produce steam in a boiler. The steam passes through a steam turbine and the exhaust steam is condensed and fed back to the boiler by a pump. There are of course many pieces of equipment in this cycle and each of them either (1) does some work like the turbines, (2) takes up some work like the pumps, (3) creates some heat like the boiler or reheater, and (4) takes up some heat like leakages. If we divide the net work done by the net heat supplied, we get the Rankine efficiency of the plant. Computing all these values requires knowledge about temperatures, pressures, and enthalpies in various parts of the plant (Eastop, 1993).

There are four main stages in the template view of a coal-fired power plant. These four steps are displayed in Figs. 11.1 and 11.2 in schematic and thermodynamic form. Water is pumped to a boiler (step 1) where it changes its phase from liquid to gas (step 2). The water vapor enters the steam turbine which converts the energy into rotational motion that generates electricity (step 3). The steam is then condensed back into liquid form and the cycle repeats. In a real power plant, there are many more pieces of equipment and the process is not ideal. The efficiency is then the net mechanical work done divided by the heat input. Usually, the exhaust steam after the steam turbine is reheated to allow a second turbine to extract more energy from the steam; in these cases, the efficiency calculation gets more terms of work output and heat input, but the basic idea of efficiency remains the same.

A gas-fired power plant is more complex in that the gas is combusted in a gas turbine that creates rotational motion, which is

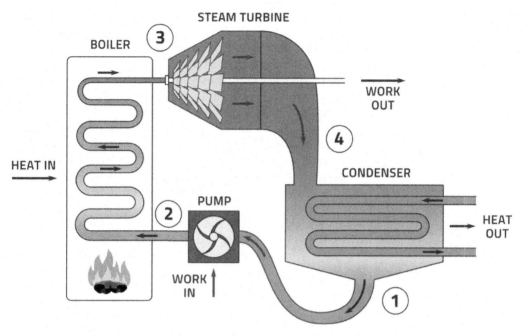

Fig. 11.1 The basic schematic of a Rankine cycle.

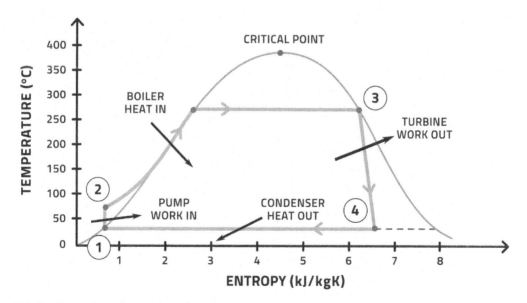

Fig. 11.2 The thermodynamic overview of the Rankine cycle.

converted to electricity. The exhaust gas from the gas turbine can then be used in a heat recovery steam generator (HRSG), which is essentially a heat exchanger that uses the hot exhaust gas to heat up water in the same way as displayed in Fig. 11.1.

This broad view can be used for almost any power plant even including solar thermal plants, with the only notable exception being photovoltaic power. All such plants are essentially based on a source of heat raising the temperature of a medium (e.g., air or water) and generating the rotation motion of a generator.

For each segment in the thermodynamic overview (Fig. 11.2), we compute the difference in enthalpy to get the energy input or output during that segment. Enthalpy cannot be measured and so we must compute it. This is not simple, but it can be done easily and accurately by using the international standard formulas of IAPWS-IF97 (Wagner and Kruse, 1998).

If we can measure process temperatures and pressures, the efficiency of a real power plant can be calculated accurately in real time with this method that can take account of all the nonidealities of a real plant.

Power plant efficiency, no matter how it is computed, changes from moment to moment. The variability of efficiency is significant even over the course of a single day. It is, for example, heavily dependent on the load of the plant, i.e., how much power it is producing. The design of the power plant considers what loads we expect the plant to have to meet during its lifetime. This assumption is made at design time before the plant is built. In 2020, over 60% of power plants in North America are over 40 years old. Most natural gas-fired plants are between 10 and 20 years old and so much younger than the coal-powered fleet.

The design of the plants built in the last 10 years or so considered that the plant load will vary often. Older plants however have been designed for a certain standard load condition that is assumed to apply most of the time. In our modern world however, even these older plants spend considerable amounts of time not at the standard load but in various lower load conditions that are less efficient. Depending on the amount of wind and solar energy produced at any one time, a fossil fuel power plant may find itself ramping its load up and down several times in 1 day, thereby significantly affecting its efficiency.

11.2 The value of efficiency

How much financial return an improvement in efficiency provides depends on the price of the fuel and electricity as well as the scale, construction cost, and operating cost of the power plant.

Simply put however, an increase in efficiency will reduce fuel consumption needed to meet a given power output. Some ways to improve efficiency require substantial capital investment or additional operating cost whereas other ways are relatively inexpensive. For example, better advanced process control technology is a comparatively small investment with potential for high return.

At the time of writing, approximately 1192 institutions controlling 14.1 trillion USD in investment have pledged to divest from fossil fuel assets (350.org, 2020). This movement is quickly growing and casts a shadow of doubt on fossil fuel power plant valuations. Assuming the plants remain in operation efficiency is the most powerful lever to affect operating income and asset value.

We will do a sample calculation of the economic value of an efficiency increase here. Let us assume that a power plant has a capacity of 600 MW but uses 15 MW internally so that its net capacity is 585 MW. The plant will not output this capacity at all times. Because electricity demand changes, the actual capacity delivered is usually less than net capacity over the course of the year. Let us say that the capacity factor is 0.25, i.e., the amount of electricity output on average as a fraction of net capacity. In our case this is $0.25 \times 585 = 146.25$ MW. The cost of fuel is usually measured in dollars per million Btu of potential energy stored in the fuel. A typical price for coal is 2 USD/MMBTU. Multiplying the effective capacity by the price and by the number of hours in 1 year, we obtain the cost of 1 Btu/kWh, which is 2562.30 USD. A typical heat rate for a coal-fired plant is 10,500 Btu/kWh in which case the cost is approximately 27 million USD per year. That is how much it costs to run a power plant, considering only the fuel cost and factoring in the internal electricity demand.

Making this power plant more efficient will allow us to use less fuel to make the same amount of electricity. A heat rate of 10,500 Btu/kWh corresponds to an efficiency of 32.5%. If we were to make this plant more efficient by 1%, the heat rate would drop to 10,185 Btu/kWh. In this case, increasing the efficiency of the plant by 1% is worth 800,000 USD per year. We note that we did not need to know the price of electricity to calculate this benefit.

Having calculated the financial benefit of efficiency gains, an investor in a power plant is primarily concerned about the margin between the cost to generate power and the revenue obtained from selling power. A typical wholesale price for power is currently about 21 USD per MWh. At the previously mentioned heat rate and coal price, the power production cost is 21 USD per MWh. It is easy to see why coal power is at cross-roads. A modern combined cycle gas-fired power plant achieves heat rates of between 6500 and 7000 and the price of gas is currently about 1.76 USD/MMBTU. This combination entails a production cost of 11.44 USD per MWh. That means that there is a margin of 9.56 USD/MWh available between the price to produce and the expected revenue. The real margin is, of course, much smaller because the plant has more costs than just its fuel.

Furthermore, we note that the heat rate is not a constant but fluctuates over time. Fig. 11.3 displays a practical case where we see the dependence of heat rate on the load of the plant over 4 years. As the plant increases its load, the heat rate can decrease by as much as 200 Btu/kWh (from 9780 to 9580 in 2016). That corresponds to an efficiency increase of 0.73%. We also see that the heat rate relative to load changes over time and tends to rise as time goes by. Over the course of 4 years, the rate of this example increased by about 250 Btu/kWh.

Heat rate degradation is typical in the first few years of a newly built power plant. This is due to many causes that may or may not be measurable or controllable. A discussion of such causes is well beyond the scope of this chapter.

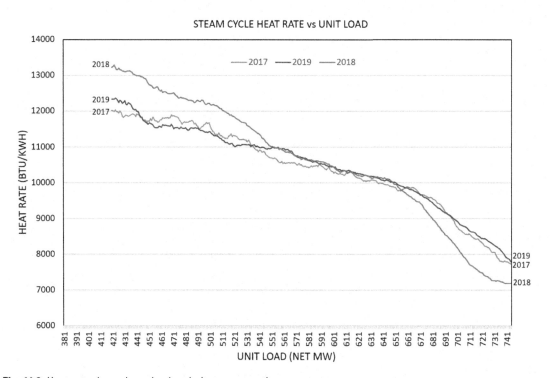

Fig. 11.3 Heat rate depends on load and changes over time.

11.3   Variable sensitivity

When faced with a phenomenon in a complex setting that is characterized by many variables, we ask which factors are causally responsible for this phenomenon. A power plant may have several thousand measured or calculated variables in its control system. Which variables are important for overall efficiency or for the efficiency of underlying processes such as boiler or combustion turbine performance? If we knew the answer to that, we could perhaps alter these variables and improve the efficiency.

A different path that can affect efficiency focuses on physical changes to the plant. For example, a pump may be exchanged for a more efficient one or the combustion process may benefit from additional burners or a pulverizer may be modified to produce a finer coal product. These possibilities are not considered here. From a financial point of view, changing set points of parameters that are automatically controlled comes at no capital cost whatsoever and usually a small operational cost. However, knowing what set point to change, when and how to change it can have significant impact on efficiency.

The act of discovering which parameters affect efficiency requires human expertise and data analysis. If a parameter known to affect efficiency is controlled automatically within the plant control system, then changing it becomes merely a matter of entering a new set point (desired) value into the control system. The small operational cost may derive from this change causing some fuel, water, air, or other consumable. If, however, the set point is not controlled automatically, we may have to install the appropriate sensor, actuator, control system element or even install novel equipment to make it available, in which case capital costs are incurred. It has been the experience of the second author that, in practice, many parameters affecting power plant efficiency are automatically controlled and therefore can be adjusted to improve efficiency in most cases.

Conversely, adjustment of some automatically controlled parameters may decrease efficiency. In one case at a combined cycle facility, the combustion turbines' firing curves were changed to increase full load output. This change led to a decrease in exhaust gas temperature which in turn caused boiler efficiency and steam turbine generator output to decrease. This case shows that seemingly small changes in one system component can affect the performance of the entire system in significant ways.

Furthermore, we note that efficiency is not usually a linear function as may have been suggested by Fig. 11.3. As we can see in Fig. 11.4, the electricity output of a steam turbine in a gas fired combined cycle power plant depends on the overall plant

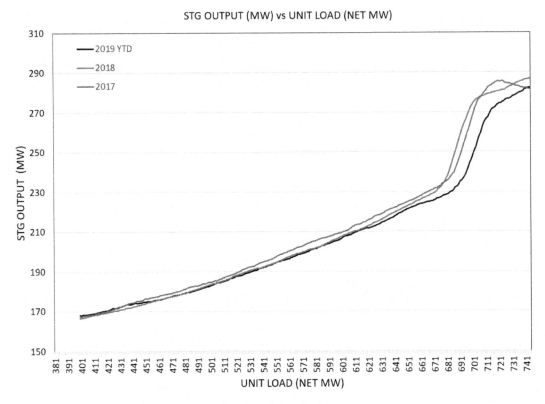

STG OUTPUT (MW) vs UNIT LOAD (NET MW)

Fig. 11.4 Steam turbine electricity output as a function of plant load.

load condition as well as many other factors. The dependence undergoes a dramatic change between about 670 and 710 MW of load because the operator turned on the duct burners. A curve shaped like the letter "S" is characteristic of what physicists call a phase transition. It is when the state of a system changes not just quantitatively but qualitatively such as the change of water when it undergoes a phase transition from liquid to steam. In this transition, water absorbs a lot of heat without raising its temperature just to effect the change in phase. A similar behavior is seen here as the turbine output is changed a lot even though plant load changes little.

The operator controls when to turn on the duct burners and has control over many other factors as well. Deciding when to change what factor is a complex task, particularly because these factors are often interrelated. Knowing which factors matter most and how they relate can offer some valuable insights into optimizing operations. Starting from the desire to maximize efficiency, the highest goal would be to have a model that could prescribe the optimum set point for all controllable parameters at any one time.

11.4 Measurability, predictability, and controllability

When considering factors affecting efficiency or heat rate, we should also consider to what extent these factors are measurable, predictable, or controllable, see Fig. 11.5. If we can measure, predict, and control a factor, then its influence is reasonably certain. For example, plant load is measurable and controllable but commonly unpredictable due to market conditions. Another example is atmospheric conditions which affect many power plant operating characteristics. These conditions are somewhat predictable but entirely uncontrollable. Whereas if we can measure, predict, and control a factor, then we can start to do something about it. Most automatically controlled parameters in a power plant control system are examples of such factors.

Some factors may not be measurable because no sensors were installed to measure them. Vibrations are a common example as vibration sensors have only recently become a quasistandard. In older plants, we find few, if any, and the installation costs are significant. Without measurement, the question of prediction and control is moot. Installing necessary instrumentation is the first step in a digitization journey.

Another difficulty is that we do not know all factors with total precision. For example, when trying to make a mathematical model of efficiency for a real power plant based on empirical data,

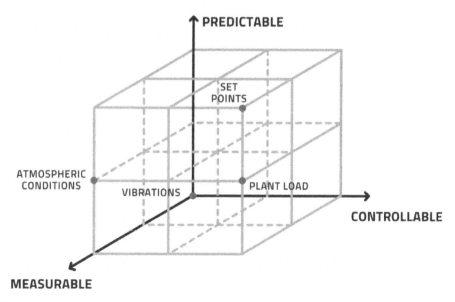

Fig. 11.5 What are the factors affecting efficiency and where are they in a matrix of predictability and controllability?

we will always end up with a calculation that differs by some amount from the true efficiency. This difference is due in part to the measurement errors of all the quantities that flow into the model and partly due to factors that we do not measure or do not know.

11.5 Process modeling and optimization

To optimize a process, we must take six basic steps, see Fig. 11.6:

(1) Identify all measurable factors (parameters) potentially affecting process performance.

(2) Quantify the **sensitivity** of process performance to changes in each individual parameter.

(3) Classify each parameter in terms of predictability and controllability.

(4) Construct a model that accurately predicts the effect of changing each controllable parameter under various combinations of uncontrollable parameters.

(5) Apply the model to the real process by changing controllable parameters in response to changes in uncontrollable parameters.

(6) Repeat Steps 1–5 continuously until no further improvement can be achieved.

Modeling is a fundamental component in the optimization sequence. The first goal of modeling must be to construct a model that is as accurate as possible. Based on this accuracy, we can then assess how much natural uncertainty there is in the system and how much is due to unknown, or unmeasured, factors.

Model accuracy depends on our ability to measure, analyze, and thereby predict the sensitivity of process performance to

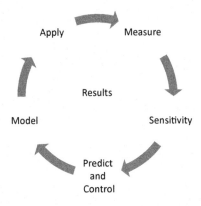

Fig. 11.6 The six steps to optimizing a process.

changes in individual parameters. Sensitivity, in general, is taken to mean by how much the system outcome changes if any one of the system inputs change. Mathematically, we can simply define this as the partial derivative

$$s_i = \frac{\partial f(x_1, x_2, \ldots, x_N)}{\partial x_i}\bigg|_{x_j = \bar{x}_j}$$

That is to say, the sensitivity of the ith factor x_i is the partial derivative of the model that depends on all its input factors with respect to the ith factor. If left like this, the sensitivity would not be a value but rather a function and so we must evaluate it at some point. The natural point is at the average value of each input factor $\bar{x}$.

While the sensitivity is an interesting number, it must be kept in mind that the reaction of the system to any factor is likely to be nonlinear. In addition, the reaction of the system to a simultaneous change in more than one factor is *not* captured by the sensitivity at all. Ranking factors by sensitivity is therefore a useful first indication of importance but not the only consideration.

How should the factors in the model be selected in the first place? In a typical power plant, thousands of parameters (such as pressure, temperature, and flow) are continuously measured and recorded (as "tags") in the plant data historian. We must choose a small number of tags for a model so that we can keep the model simple enough to produce a reliable prediction. Also, we do not want to present data to the model that has nothing to do with the effect under question. The best way to choose relevant tags is through expert domain knowledge. If we are uncertain about the influence of a tag, it is probably best to include it and see what its sensitivity is. If the sensitivity is exceptionally low, we can probably ignore it.

Another way to identify relevant tags is by performing a correlation analysis. First, the matrix of correlation coefficients between every pair of available tags is calculated. Second, we select all those tags that highly correlate with our quantity of interest. Third, we look at the pairwise correlations of all tags on this list. If any pair correlates very highly with each other, we keep only one of them. In this way, we will get a relatively small list of tags that correlate with the outcome and do not correlate too much with each other. That should be a reasonable start for the list of relevant tags and usually captures most of the variance in the available data.

Adding domain knowledge to a completed correlation analysis usually leads to a model that captures about 90% of the variance and this is often the best that we can expect under realistic power plant operating conditions.

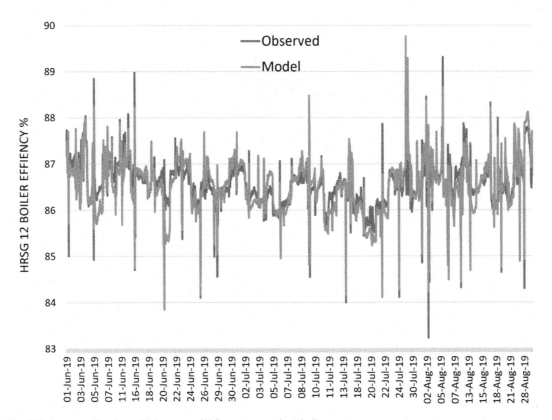

Fig. 11.7 An example of an efficiency model for a power plant boiler.

An example of such a model can be seen in Fig. 11.7 in which boiler efficiency (steam energy produced/flue gas heat energy absorbed) predicted by the model compares well with actual data. In this case the model was produced by machine learning analysis of process data which yielded five parameter tags correlating with efficiency:

- HRH ATTEMP APPROACH TEMP (DEG F): Difference between dry bulb temperature and saturation temperature of steam at the exit of the reheat attemperator section.
- RH ATTEMP OUTLET STEAM TEMP (°F): Dry bulb temperature of steam at the exit of the reheat attemperator section.
- IP STEAM HDR FLOW (KPPH): Mass flowrate of steam between the intermediate pressure sections of the boiler and turbine.
- RH ATTEMP OUTLET DRN TEMP A (°F): Dry bulb temperature of condensate in the drain lines downstream of the reheat attemperator section.

- CALCULATED HRSG12 HRH FLOW: The calculated value (not directly measured but deduced from other instrumentation) of steam mass flowrate in the reheat section of the boiler.

One factor (RH ATTEMP OUTLET STEAM TEMP) from the previously mentioned list is controllable and several others are directly related to it; therefore, efficiency may be improved by changing the controllable factor in response to variation of uncontrollable factors in accordance with the model.

References

350.org, 2020. Retrieved April 2020. https://gofossilfree.org/divestment/commitments/.

California Energy Commission, 2019. Updating thermal power plant efficiency measures and operational characteristics for production cost modeling. CEC-200-2019-001. California Energy Commission.

Eastop, T.D., 1993. Applied Thermodynamics for Engineering Technologists. Longman, Harlow.

Wagner, W., Kruse, A., 1998. Properties of Water and Steam. Springer, Heidelberg.

12

Forecasting wind power plant failures

Daniel Brenner[a], Dietmar Tilch[b], and Patrick Bangert[c,d]

[a]Weidmüller Monitoring Systems GmbH, Dresden, Germany
[b]ZF Friedrichshafen AG, Lohr am Main, Germany
[c]Artificial Intelligence, Samsung SDSA, San Jose, CA, United States
[d]Algorithmica Technologies GmbH, Bad Nauheim, Germany

12.1   Introduction

Many problems of technical equipment manifest in the vibration behavior at certain locations. If a vibration sensor is put there and the vibration is measured, then we can ask if we could identify the problem before it leads to any actual damage. This chapter will discuss how to establish correlations between vibration measurements and technical problems so that a reaction is possible before the damage occurs. This will be done using the example of rotor blades on wind power plants. Specifically, a large glued structure part unglues quickly from the rotor hull and causes a significant damage that could lead to total loss of the blade. The prediction system here described can also predict damages in bearings, the gearing mechanism, the generator, the oil circulation, and other components of a wind power plant.

Machine Learning and Data Science in the Power Generation Industry. https://doi.org/10.1016/B978-0-12-819742-4.00012-3
© 2021 Elsevier Inc. All rights reserved.

The essential problem of maintenance is its reactive paradigm. One waits until something happens and then repairs it. Therefore one always works under time pressure. The customer does not only pay for the (1) necessary repair costs but also the (2) collateral damage that often amounts to several times the repair cost. Of course, the customer must also accept the cost of (3) loss of production. Additionally, the factor of (4) inability to plan the repair work must be added: spare parts must be kept on stock or ordered in a hurry, personnel must be rushed to the power plant (off-shore this is a significant cost), and repairs done in haste are generally poorly planned and represent a higher risk for the future, etc.

A damage prediction represents the possibility to make a paradigm shift to preventative maintenance. If one knows today that a plant will fail in 7 days, then the repair can be planned today, and the plant shut down normally in the next days and repaired. The customer only pays the normal repair costs. The collateral damage does not occur at all because the failure event itself never occurs. The loss of production is limited to the normal repair time and is thus much smaller. The factors relating to the inability to plan are significantly reduced or entirely removed.

12.2 Impact of damages on the wind power market

Wind power plant experience failures that lead to financial losses due to a variety of causes. Please see Fig. 12.1 for an overview of the causes, Fig. 12.2 for their effect, and Fig. 12.3 for the consequences implemented. Fig. 12.4 shows the mean time between failures, Fig. 12.5 the failure rate per age, and Fig. 12.6 the shutdown duration and failure frequency.[a] From these statistics we may conclude the following:

- At least 62.9% of all failure causes are internal engineering related failure modes while the remainder are due to external effects, mostly weather related.
- At least 69.5% of all failure consequences lead to less or no power being produced while the remainder leads to aging in some form.
- About 82.5% of all maintenance activity is hardware related and thus means logistical arrangements being made.
- On average, a failure will occur once per year for plants with less than 500 kW, twice per year for plants between 500 and 999 kW, and 3.5 times per year for plants with more than 1 MW of power output. The more power producing capacity a plant has, the more often it will fail.

[a]All statistics used in Figs. 12.1–12.6 were obtained from ISET and IWET.

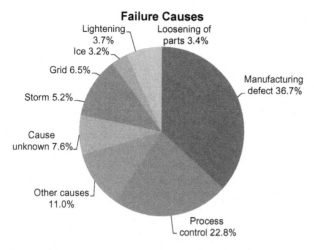

Fig. 12.1 The causes for a wind power plant to fail are illustrated here with their corresponding likelihood of happening relative to each other.

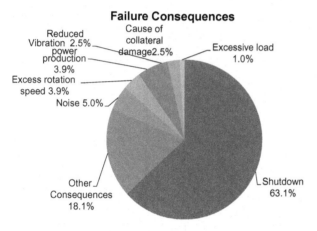

Fig. 12.2 The effects of the causes of Fig. 12.1 are presented here with the likelihood of happening relative to each other.

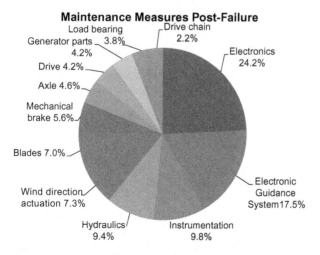

Fig. 12.3 The maintenance measures put into place to remedy the effects of Fig. 12.2 with the likelihood of being implemented relative to each other.

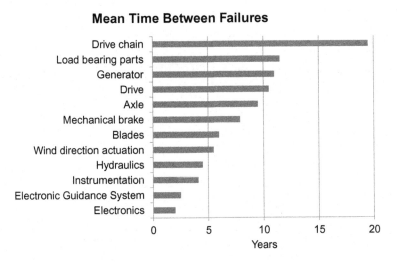

Fig. 12.4 The mean time between failures per major failure mode.

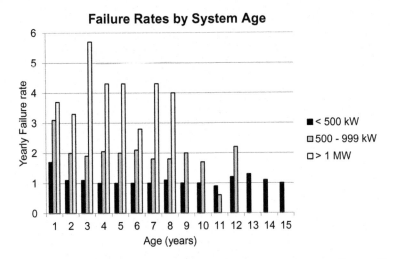

Fig. 12.5 The yearly failure rate as a function of wind power plant age. Plants with higher output fail more often and that age does not significantly influence the failure rate.

- The age of a plant does not lead to a significantly higher failure rate.
- The rarer the failure mode, the longer the resulting shutdown.
- A failure will, on average, lead to a shutdown lasting about 6 days.

From this evidence, we must conclude that internal causes are responsible for a 1% capacity loss for plants with less than 500 kW, 2% for plants between 500 and 999 kW, and 3.5% for plants with more than 1 MW of power output.

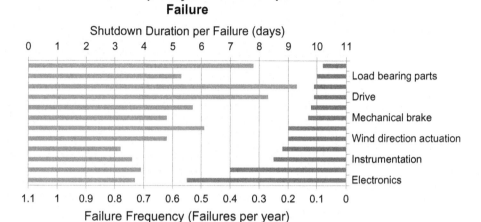

Fig. 12.6 The failure frequency per failure mode and the corresponding duration of the shutdown in days.

In a wind power field like Alpha Ventus in the North Sea, with 60 MW installed and expecting 220 GWh (i.e., an expectation that the field will operate 41.8% of the time) of electricity production per year, the 3.5% loss indicates a loss of 7.7 GWh. At the rate of German government regulation of 7.6 Eurocents per kWh, this loss is worth 0.6 million Euro per year. Every cause leads to some damage that usually leads to collateral damages as well. Adding the cost of the maintenance measures related to these collateral damages themselves yields a financial damage of well over 1 million Euro per year. The original cause exists and cannot be prevented but if it could be identified in advance, then these associated costs could be saved.

This calculation does not consider worst-case scenarios such as the plant burning up and thus effectively requiring a new build.

12.3 Vibration spectra

A vibration measurement is usually a vibration spectrum. One defines a frequency corridor $[f_a, f_b]$ and divides it into equally sized bins. As spectral analysis makes frequent use of the Fourier transform and therefore the FFT algorithm, it is opportune to select the number of bins as a power of two. A normal value for the number of bins is $2^{10} = 1024$, $2^{11} = 2048$, $2^{12} = 4096$, or $2^{13} = 8192$. A lower value would offer too little resolution and a higher value would lead to an exceptionally large data size.

We now measure the strength of the vibration, called the amplitude, for every frequency. Spectral analysis usually assumes that the time difference between any two neighboring spectra is the same. It is unreasonably effortful to ignore this assumption. How often one should measure a spectrum depends on the speed of the damage mechanism: We must decide on a case-by-case basis.

From the analysis of emission spectra of chemical elements, we know that one can conclude based on the presence/absence of spectral lines, their height, width, and combination what the chemical composition of the sample is with high precision. This analysis cannot be fully automated at current state of the art, i.e., one must tell the computer certain limiting assumptions, e.g., that certain elements are assumed to be absent. However, a largely automated detection is possible. This is based on (1) mathematical data cleaning, (2) mathematical computation of spectral line form (height, width, area), and (3) expert knowledge about which elements give rise to which lines.

The analysis of vibration spectra is more complicated due to two factors. First, the measurement is much less precise than an optical emission spectrum mostly because of noise. Vibrations are necessarily measured during operations under all possible conditions whereas optical measurements are usually made under strictly controlled laboratory conditions. Second, enough expert knowledge is generally missing about the correlation between vibrations and problems or normal behavior. Even the concept of "normal behavior" is a collection of a wide variety of operational conditions all of which vibrate differently. In addition, the vibration behavior changes due to the make, model, and age of the equipment.

Is it even possible? Yes. In the following, we show using an example from a wind power plant how it is possible.

12.4 Denoising a spectrum

Initially, we collect spectra over a period that consists of various normal conditions and diverse interesting problem conditions. These spectra must now be cleaned by removing noise and background vibrations. There are principally two ways to do this.

The first way to denoise a spectrum is the mathematical and easier way. The spectra are filtered. That is they are transformed into the frequency domain using FFT, the low frequencies (high band pass filter) and high frequencies (low band pass filter) are deleted, the remaining frequencies are multiplied by a factor $w < 1$ (Wiener filter), and the spectrum is transformed back into the time domain. How many low and high frequencies must be deleted depends on the equipment and its environment. One can estimate this visually by looking at several transformed spectra from normal operations and looking for the frequencies at the edge of the spectrum where the frequencies are overrepresented. Those frequencies may be deleted. The Wiener filter is a general noise filter based on the assumption that the noise is normally distributed. This is, of course, never the case and that is why we also employ the other two filters. The factor w is the estimated relationship between the amplitude of the desired signal to the total amplitude: Is the signal approximately 85% of the measurement and the other 15% are noise, then $w = 0.85$. This factor is normally estimated visually using the spectra from normal operations.

The second way to denoise a spectrum is to know the background noises, i.e., the spectrum of the noise, and to subtract it from the spectrum. For this to work, we must however install a second vibration measurement that measures the background noise. The subtraction may then be made in situ or later in an analysis phase. As not all noise can be eliminated in this way, the previous mathematical analysis should also be applied, albeit with different values.

Of course, the second way is better. However, it requires an investment in hardware, software, and effort. If it is installed later, we must wait for some time before enough data is available to be analyzed. Depending on how often the problems occur, this may take a long time.

Is it worth it? Yes. In Fig. 12.7 we show the difference between a raw spectrum and a filtered spectrum according to the mathematical method. One can now reliably compute the height, width, and area of a spectral line.

12.5 Properties of a spectrum

Due to noise, the height of a spectral line is very imprecisely measured. This means that the absence of a line (height near zero) cannot be empirically determined with certainty. Even an extremely high line does not have a precisely determinable height. The line height thus has a large uncertainty that must be determined empirically. We determine this uncertainty by comparing many technically identical normal operational conditions assuming that the line height was actually the same but merely measured to be different. This results in a distribution of heights around an average. This distribution should roughly be normal in shape and then we may assign the uncertainty a factor multiplied by the standard deviation of this distribution. A factor of 2 or 3 is sensible as one can include nearly all cases in this way. The factor must be chosen pragmatically so that the uncertainty does not become so large that the line height becomes wholly indeterminable.

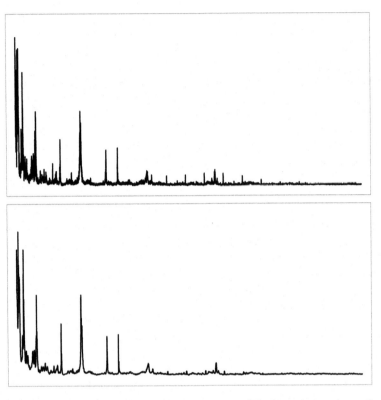

Fig. 12.7 The upper image shows a raw spectrum and the lower image shows the same spectrum filtered as described in the text. The *sharper lines* and the reduced noise are clearly visible.

From the spectrum in Fig. 12.7, we can see nine lines clearly and can determine their heights, widths, and area as well as their locations. Is the height larger than a minimum height, we say that the line is "present" and otherwise "absent." The area is generally determinable as a function of height and width and therefore is not really an independent property. We therefore have three interesting values per line (presence, width, and height). The full spectrum of 8192 numerical entries is thus compressed to 3 properties of 9 lines, i.e., 27 numbers.

With optical spectra, one would now rely on tables of known spectral lines to determine the substances that make up the material at hand. In our case, such information is missing.

12.6 Spectral evolution

As the correlation between damage mechanism and spectral lines depends on make and model as well as age of the equipment, we will not even try to determine the missing expert knowledge in the usual table format as this would require immense effort. Instead, we will use time series analysis from the area of machine learning.

We do not want to analyze any one spectrum but rather the totality of spectra taken over a period that is as long as possible. We extract the properties as discussed from each individual spectrum and thus obtain an evolution of properties over time. Individual lines appear, disappear, becomes stronger or broader and weaker. The evolution can be presented as a spectrogram, see Fig. 12.8.

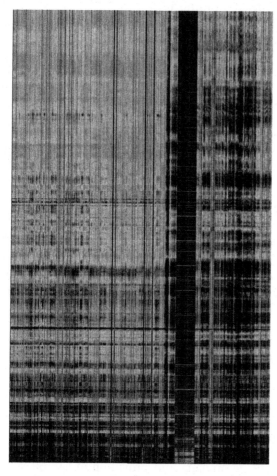

Fig. 12.8 A spectrogram with the frequencies on the vertical axis and time on the horizontal axis with the whole axis denoting about 6 months. The colors denote the amplitudes. It is easy to see a shutdown phase after about two-thirds of the time and a significant increase in vibration in the time leading up to this shutdown.

In addition to the empirical knowledge, we collect the times of all the occurred damage events in the relevant time period and classify them according to their type.

What we need now is a numerical variable, for each damage type, that can be computed using only the spectral properties and that exceeds a limiting value if and only if the damage mechanism is present. This is usually achievable by a linear function with appropriate coefficients, i.e., linear regression line. As this is a standard mathematics exercise it will not be discussed here.

The key is to predict the damage signals into the future, i.e., to say today that the day after tomorrow the limiting value will be reached and that therefore a damage will result and thus the plant should be turned off tomorrow.

12.7 Prediction

For prediction, we now have a single signal in time for each damage mechanism, computed from the properties of the spectra. This is an essential compression of the empirical data. Various measurement problems are thus averaged out of the data and the result is stable and robust. A damage mechanism always has a "memory," i.e., the current state does not only depend upon the immediately preceding state but rather upon a history. This means that time plays an essential role in prediction.

As we are going to use machine learning due to pragmatic constraints and wish to predict a time series that depends upon a longer history of itself, it is sensible to use recurrent neural networks (Bangert, 2012).

Research shows excellent results in the use of recurrent neural networks, especially the long short-term memory network, in the prediction of diverse time series. Even after decades of research into neural networks in general and several years of research into recurrent neural networks specifically, the parameterization of the topology and learning algorithms continues to be an art form that can only be mastered by experience. Is the network trained, either by experience or trial and error, it will yield reliable predictions for the time series on which it was trained (Mandic and Chambers, 2001).

If it requires such an effort, why is it called machine learning? Like human learning that usually requires a teacher, machine learning also requires a teacher. The teaching activity consists of the setting of parameters and providing of sample output data. The actual learning then takes place automatically (Bishop, 2006).

Recurrent neural networks have the positive property that they can be updated. If we gain new knowledge from further measurements, this knowledge can be incorporated into the network without necessitating a full training. Thus this updating can occur fully automatically with the same parameterization as the original training. The network is therefore always up to date without human effort.

12.8 Results on turbine blades

For this study, we investigated four wind power plants. Each plant experienced exactly one fault in the chosen period and all four faults were of the same type, i.e., the ungluing of the structure part in the blade. We knew when two of the faults occurred. Using these two plants' datasets, the model was constructed and then applied to the other two plants. The model was able to correctly predict the other two faults approximately 5 days before they happened.

Except for the vibration spectra, no other data was used. The analysts had the information when the fault occurred on two of the plants. No other information about the nature of the fault was known.

The prediction is correct and reliable in that it correctly predicted the known faults 5 days before they happened. Furthermore, no nonexistent faults were predicted. The method is therefore practically usable. The major advantage of this method over conventional methods is the high degree of automation after the initial parameterization as the model keeps itself up to date and delivers fully automated predictions. The paradigm shift from reactive to preventative maintenance can now happen and then owner of the plant can save unnecessary costs.

12.9 Results on the rotor and generator

A similar method was applied to the data from the rotor and generator of a wind turbine to examine the damages that can occur to the core of the power plant as opposed to the blades examined previously. From the instrumentation, all values were recorded to a data archive for 6 months. One value per second was taken and recorded if it different significantly from the previously recorded value. A total of 56 measurements were available from around the turbine and generator but also subsidiary systems such the lubrication pump. Using 5 months of these time series, a model was created and found that the model agreed with the last month of experimental data to within 0.1%. Thus we can assume that the model correctly represents the dynamics of the wind power plant.

It should be noted that the model does not model each time series separately but rather models all time series as an integral whole. This is one of the main benefits of using this type of modeling approach in that all dependencies are accounted for.

This system was then allowed to make predictions for the future state of the plant. The prediction, according to the model's own calculations, was sometimes accurate up to 1 week in advance. Naturally, such predictions assume that the conditions present do not change significantly during this projection. If they do, then a new prediction is immediately made. Thus if, for example, a storm suddenly arises, the prediction must be adjusted.

One prediction made is shown in Fig. 12.9, where we can see that a vibration on the turbine is to exceed the maximum allowed alarm limit after 59 ± 5 h from the present moment. Please note that this prediction means that the failure event will take place somewhere in the time range from 54 to 64 h from now. A narrower range will become available as the event comes closer in time. This information is, however, accurate enough to become actionable. We may schedule a maintenance activity in 2 days from now that will prevent the problem. Planning for 2 days in advance is sufficiently practical that this would solve the problem in practice.

In this case, no maintenance activity was performed in order to test the system. It was found that the turbine failed due to this

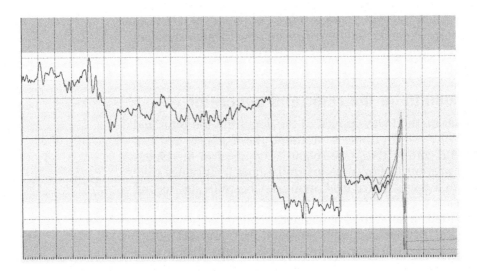

Fig. 12.9 The prediction for one of the wind power plant's vibration sensors on the turbine clearly indicating a failure due to excessive vibration. The *vertical line* on the last fifth of the image is the current moment. The *curve* to the left of this is the actual measurement, the *curve* to the right shows the model's output with the confidence of the model.

vibration exceeding the limit after 62 h from the moment it was predicted to happen. This failure led to contact with the casing, which led to a fire effectively destroying the plant.

It would have been impossible to predict this event more than 59 h ahead of time due to the qualitative change in the system (the failure mode) occurring a few days before the event. The model must be able to see some qualitative change for some period before it is capable of extrapolating a failure and so the model has a reaction time. Events that are caused quickly are thus predicted relatively close to the deadline. In general, failure modes that are slower can be predicted longer in advance.

It must be emphasized here that the model can only predict an event. It cannot label this event with the words "will cause a generator fire." The identification of an event as a certain type of event is another matter. It is possible via the same sort of methods but would require several examples of each failure mode and this is a practical difficulty. Thus the model can give a specific time when the system will suffer a major defect; the nature of the defect must be discovered by manual search on the physical system.

References

Bangert, P., 2012. Optimization for Industrial Problems. Springer, Heidelberg.

Bishop, C.M., 2006. Pattern Recognition and Machine Learning. Springer, Heidelberg.

Mandic, D., Chambers, J., 2001. Recurrent Neural Networks for Prediction: Learning Algorithms, Architectures and Stability. Wiley, Hoboken, NJ.

Index

Note: Page numbers followed by *f* indicate figures and *t* indicate tables.

Printed in the United States
By Bookmasters